Guide to 'H' Grade Human Biology Essays

by

John Di Mambro
and
Mark Dorris

© *J. Di Mambro and M. Dorris, 1994*
ISBN 0 7169 3188 5

ROBERT GIBSON · Publisher
17 Fitzroy Place, Glasgow, G3 7SF.

INTRODUCTION

ADVICE TO CANDIDATES

1. Each relevant point commences with a capital letter and related points are indented.

2. In the 'Relevant Points' for each question, points which could be substituted by the use of appropriately labelled diagrams are marked by a bracket running down the left-hand margin.

3. Where a double-space occurs in a list of points, this indicates the start of a new paragraph.

4. The Topic (in capitals) and Sub-Topic (in lower case) from the Scottish Examination Board's syllabus for the Higher Grade Human Biology have been given in the Contents for each essay. This should enable you to easily locate an essay for a particular section of the syllabus.

5. Answers can be improved by relevant diagrams, formulae or equations. If you inadvertently write something which is ambiguous, a diagram, formula or equation may make clear that you really understand the point.

6. As a guide for candidates, two specimen answers based on the appropriate list of 'Relevant Points' are included, these are for Essay Questions 1 and 2. For each essay, one specimen answer is purely textual while the other achieves a similar standard with the use of appropriately labelled diagrams.

7. These points to be covered form the skeleton of the answers required. It will be for you, the candidate, to 'flesh' these out using properly constructed sentences and paragraphs. Where appropriate, show connections between these by means of link words such as 'therefore', 'because' and 'in order to'.

8. A common problem when writing an essay is knowing how to begin. An introductory paragraph can be useful in setting out some of the major points. By way of illustration, every essay has such an introduction.

9. A closing paragraph in the form of a brief discussion may also be helpful in summing up some of the main points.

10. Before starting your answer, make a short plan. Candidates often feel they cannot spare this time, yet experience shows that a few minutes spent here will result in a better constructed essay. Such a plan reduces

the risk of omitting relevant facts and ideas or omitting them from opening paragraphs and have to add them in paragraphs where they do not belong.

11. Be careful to pace yourself in this section of the examination. There is a law of diminishing return for the time you spend on a single question — so spread your effort over both questions, paying particular attention to the allocation of marks if the essay is in parts. If you were to devote all your time to just one essay, even with a perfect answer, you could only score fifteen marks out of thirty. It is easier to pick up more marks by doing both essays.

12. In general, an answer should cover at least as many points as there are marks allocated for the question.

13. Your information must be relevant. The inclusion of irrelevant information, even if it is itself correct, will gain you no marks.

14. The points listed for inclusion in essay answers to the following questions are not the total of all information on each subject and you may be able to add facts and ideas from your own knowledge. If these are correct and relevant, you will be credited accordingly. This is not necessary, however. High scores would be attainable even if some of the points listed here were omitted and a good essay composed from the remainder.

15. A good scientific essay is a combination of knowledge and expression requiring practice. You can hope to improve only by repeated attempts, thus building up your confidence, timing and expertise.

COPYING PROHIBITED

Note: This publication is NOT licensed for copying under the Copyright Licensing Agency's Scheme, to which Robert Gibson & Sons are not party.

All rights reserved. No part of this publication may be reproduced; stored in a retrieval system; or transmitted in any form or by any means — electronic, mechanical, photocopying, or otherwise — without prior permission of the publisher Robert Gibson & Sons, Ltd., 17 Fitzroy Place, Glasgow, G3 7SF.

CONTENTS

SPECIMEN ANSWERS

1. Outline the structure and composition of nucleic acids under the following headings:
   (a) differences between DNA and RNA;
   (b) structure of DNA in relation to its replication. 10

Transport mechanisms — the need for a transport system/tissue fluid and lymph

2. Give an account of the transport system under the following headings:
   (a) need for a transport system;
   (b) basic organisation;
   (c) formation of tissue fluid. 15

CELL FUNCTION

Cell metabolism — the role of enzymes

3. Discuss the role of enzymes as catalysts in biochemical pathways. 20
4. Describe the factors which determine enzyme activity under the following headings:
   (a) inhibition;
   (b) concentration of enzyme and substrate. 22
5. Functional enzymes are essential in normal metabolism. Discuss this statement under the following headings:
   (a) activation of enzymes;
   (b) absence of particular enzymes. 24

Cell metabolism — protein synthesis

6. Proteins have many diverse roles in living systems. Discuss this statement under the following headings:
   (a) intracellular proteins;
   (b) proteins outwith the cell. 26

Cell metabolism — energy transfer

7. Discuss the importance of energy being released in a controlled way. 28
8. The body sometimes utilises compounds other than glucose as sources of energy. Discuss this statement under the following headings:
   (a) marathon running;
   (b) starvation. 30

Cell transport — membranes

9. Cell membranes are fluid and porous in nature. Discuss why this is important in relation to the function of each of the following:
   (a) plasma membrane;
   (b) rough endoplasmic reticulum;
   (c) Golgi apparatus. 32
10. Describe the functions of lipids and proteins within the plasma membrane. 34

Cell transport — diffusion and osmosis/active transport
11. The need for a constant environment within the cell is essential in living systems. Discuss this statement. 36

Cellular response in defence — recognition of antigens
12. Using an example you have studied, write an essay on recognition of antigens. 38

Cellular response in defence — production of antibodies
13. Distinguish between cellular and humoral responses as defence mechanisms. 40

Cellular response in defence — function of macrophages
14. Describe the function of macrophages. 42

Cellular response in defence — immunity
15. Describe the two forms of acquired immunity. 44
16. Discuss immunity under the following headings:
    (a) autoimmunity;
    (b) allergy. 46

Cellular response in defence — the nature of viruses and their invasion of cells
17. What is the nature of viruses and how do they invade cells? 48
18. By means of an example you have studied, outline the life cycle of a virus and its effects on the host cell. 50

CONTINUATION OF THE SPECIES

Inheritance — chromosomes as vehicles of inheritance
19. Compare and contrast mitosis and meiosis. 52
20. Describe the factors leading to variation under the following headings:
    (a) independent assortment;
    (b) crossing-over;
    (c) mutation. 54

Inheritance — monohybrid inheritance
21. Discuss inheritance under the following headings:
    (a) patterns of dominance;
    (b) multiple alleles. 56
22. Describe monohybrid inheritance in humans. 58
23. With named examples, discuss how the inheritance of sex-linked traits differs from that of autosomal traits. 60

Inheritance — mutations and chromosome abnormalities
24. Discuss how genetic abnormalities can result in certain human conditions. 62

Inheritance — genetic screening and counselling
25. By means of examples you have studied, discuss genetic conditions of medical importance with reference to the following:
    (a) family history;
    (b) use of karyotypes;
    (c) post-natal screening. 64

Reproduction — hormonal control
26. How is the menstrual cycle controlled by hormones? 66

Reproduction — intervention in fertility
27. Discuss fertility intervention with reference to the following:
    (a) causes of infertility;
    (b) treatment of infertility. 68

Development — intra-uterine development
28. Discuss intra-uterine development under the following headings:
    (a) exchange between maternal and foetal circulations;
    (b) influence of placental hormones. 70

Development — birth
29. Discuss birth with reference to each of the following headings:
    (a) induction of birth;
    (b) nutrition of the newborn. 72

Development — the pattern of growth after birth
30. Describe the pattern of growth after birth under the following headings:
    (a) major stages of growth;
    (b) role of growth hormones;
    (c) hormonal changes and development at puberty. 74

LIFE SUPPORT MECHANISMS

Transport mechanisms — the need to circulate fluid in vessels
31. Describe in detail the structures through which oxygen and carbon dioxide will pass as they are transported between the alveoli in the lungs and respiring cells in the brain. 76
32. How is blood pressure in vessels caused and how does it change in different parts of the circulation? 77
33. As well as circulating fluid in blood vessels, the body circulates fluid in lymph vessels. Give an account of this 'secondary circulatory system' under the following headings:
    (a) formation and importance of lymph;
    (b) lymph circulation;
    (c) role of lymph nodes in defence. 79

Delivery of materials to cells — oxygen

34. Red blood cells are amongst the most unusual and plentiful cells in the human body. Write an account of these cells with reference to the following:
    (a) relationship between structure and function;
    (b) production and eventual breakdown. 81

Removal of materials from the blood — the role of the liver

35. Discuss the role of the liver under the following headings:
    (a) metabolism of protein;
    (b) detoxification;
    (c) conservation of useful substances. 83

Removal of materials from the blood — the role of the kidneys

36. Give an account of the role of the kidneys with reference to the following:
    (a) ultrafiltration;
    (b) reabsorption. 85

Regulating mechanisms

37. With reference to the maintenance of blood sugar level, explain what is meant by negative feedback control. 87
38. Discuss physiological regulation with reference to the following:
    (a) control of heart rate;
    (b) effects of exercise on the cardiovascular system. 89
39. The temperature of the body is normally very stable despite fluctuations in the environment. How is this stability achieved? 91

THE BIOLOGICAL BASIS OF BEHAVIOUR

Nervous system and memory — the brain

40. Give an account of functional areas in the cerebrum. 93

Nervous system and memory — organisation of the nervous system

41. Contrast the two divisions of the autonomic nervous system. 95
42. Discuss the role of the synapse in the transmission of nerve impulses. 97
43. Outline the organisation of the nervous system under the following headings:
    (a) central nervous system;
    (b) consequences of myelination;
    (c) plasticity of the response of the nervous system. 99

Nervous system and memory — memory

44. Learning depends heavily on information being encoded, stored and retrieved. How are each of these processes thought to occur? 102
45. Outline the evidence for the molecular basis of memory. 104

Factors influencing the development of behaviour — maturation

46. By means of a suitable example, show how maturation can influence the development of behaviour. 106

Factors influencing the development of behaviour — inheritance
47. Discuss the role of inheritance in behavioural development. Your answer should include information about:
    (a) role of the genotype;
    (b) a selected inherited disorder. 108

Factors influencing the development of behaviour — environment / the inter-relationship between maturation, inheritance and the environment.
48. How does the development of intelligence demonstrate the influence of a combination of factors? Your answer should include information about:
    (a) maturation;
    (b) polygenic inheritance;
    (c) environment. 110

Communication and social behaviour — the effect of infant attachment
49. Why is the period of infant dependence so long in humans? 112

Communication and social behaviour — the effect of communication
50. Why is non-verbal communication so important in adult communication?. 114
51. Language allows the transfer of information, thus accelerating learning and intellectual development. Discuss this statement. 116

Communication and social behaviour — the effect of experience
52. How does repeated use of a motor skill result in the establishment of a motor pathway? 118
53. Discuss human behaviour under the following headings:
    (a) imitation;
    (b) reinforcement;
    (c) shaping;
    (d) extinction. 120
54. By means of suitable examples, distinguish between generalisation and discrimination as these apply to human behaviour. 122

Communication and social behaviour — the effect of group behaviour and social influence
55. People may show anti-social behaviour when with a group but not when on their own. How can this be explained? 124
56. Discuss the strategies used to alter human attitudes under the following headings:
    (a) internalisation;
    (b) identification. 126

POPULATION GROWTH AND THE ENVIRONMENT
Population change — human population growth
57. Discuss human population growth under the following headings:
    (a) modern human pre-history;
    (b) growth to the 20th century. 128

Population change — population control
58. By means of suitable examples, discuss factors which can control human population growth. 130

Population limiting factors — food supply
59. Scientific advances have resulted in more efficient food production. Discuss this statement. 132

Population limiting factors — water supply.
60. What are some of the problems of providing clean water and how can these be solved? 134

Population limiting facors — disease
61. Discuss disease under the following headings:
    (a) regulatory effects on populations;
    (b) use of vaccines. 136

Population effects on the environment — disruption of food webs
62. Discuss the effect of population growth on the disruption of food webs. 138

Population effects on the environment — disruption of nitrogen cycle.
63. Disruption of essential nutrient cycles can have serious environmental consequences. Discuss this statement with respect to the nitrogen cycle. 141

Population effects on the environment — disruption of carbon cycle
64. Discuss some of the possible detrimental effects on the environment of increasing the levels of atmospheric carbon compounds. 143

1. OUTLINE THE STRUCTURE AND COMPOSITION OF NUCLEIC ACIDS UNDER THE FOLLOWING HEADINGS:

(a) differences between DNA and RNA; (6)

(b) structure of DNA in relation to its replication. (9)

Introduction

Nucleic acids are stable molecules which enable living organisms to reproduce their entire genetic code from one generation to the next. They also allow for all subsequent proteins necessary for survival to be made as required.

RELEVANT POINTS

(a) Differences between DNA and RNA

The differences between DNA and RNA are significant in terms of their respective functions.

- DNA contains the 5-carbon sugar deoxyribose,
  - RNA contains the 5-carbon sugar ribose.
- DNA contains the four nitrogenous bases,
  - Cytosine, guanine, thymine and adenine,
  - RNA contains the base uracil instead of thymine.
- DNA is a double-stranded polymer,
  - RNA is a single-stranded polymer.

DNA exists usually as one type of molecule,
  RNA can exist as different types,
  Such as messenger and transfer.
DNA is a very large molecule,
  RNA is a relatively smaller molecule.

(b) Structure of DNA in relation to its replication

- Basic structural unit of DNA is the nucleotide,
  - Formed from one of the four nitrogenous bases,
  - Which pair in a complementary fashion,
  - Adenine with thymine,
  - Cytosine with guanine,
  - A sugar,
  - And a phosphate group.

Sequence of nucleotides is the primary structure of DNA.
Sugar and phosphate form the backbone of the molecule,
    While the complementary bases are located on the inside,
    And held together by weak hydrogen bonding.

Nucleotides are linked to form two chains,
    Which are complementary to each other,
    And form a double helix,
    In which two helices are wound round each other.
The helices are anti-parallel,
    Their sequences running in opposite directions.

Linear arrangement of the bases in the structure,
    Is the basis of the genetic code,
    Having enough redundancy,
    To code for all natural amino acids.

Replication of DNA starts with a breakage in the weak hydrogen bonds,
    Between the complementary bases,
    Exposing two naked single strands,
    Each of which effectively acts as a template,
    Allowing complementation of the freely available nucleotides,
    To begin building up new daughter DNA molecules,
    Which take up the same helical structure,
    As the original parental DNA molecule.

SPECIMEN ANSWERS — QUESTION 1

EXAMPLE 1

Nucleic acids are stable molecules which enable living organisms to reproduce their entire genetic code from one generation to the next. They also allow for all subsequent proteins necessary for survival to be made as required.

(a) Differences between DNA and RNA

The differences between DNA and RNA are significant in terms of their respective functions. DNA contains the 5-carbon sugar deoxyribose while RNA contains the 5-carbon sugar ribose. DNA contains the four nitrogenous bases cytosine, guanine, thymine and adenine while RNA contains the base uracil instead of thymine. DNA is a double-stranded polymer while RNA is a single-stranded polymer. DNA exists as one type of molecule while RNA can exist as different types such as messenger and transfer RNA. DNA is a very large molecule while RNA is a relatively smaller molecule.

(b) Structure of DNA in relation to its replication

The basic structural unit of DNA is the nucleotide formed from one of the four nitrogenous bases, a sugar and a phosphate group. The bases pair in a complementary fashion, adenine with thymine, cytosine with guanine.

The sequence of the nucleotides is the primary structure of DNA. Sugar and phosphate molecules form the backbone of the molecule while the complementary bases are located on the inside and held together by weak hydrogen bonding.

Nucleotides are linked together to form two chains which are also complementary to each other. The two chains form a double helix in which two helices are wound round each other. The helices are anti-parallel, that is, their sequences run in opposite directions.

The basis of the genetic code is the linear sequence of the bases in the DNA structure. This sequence has enough redundancy to code for all the naturally occurring amino acids.

DNA replication starts with a breakage in the weak hydrogen bonds between complementary bases. This exposes two naked single strands, each of which effectively can act as a template to build up new daughter DNA molecules by complementary pairing with the freely available nucleotides. Thus it is that the new helices formed are the same as the original parental DNA molecule.

EXAMPLE 2 (with a diagram)

Introduction

Nucleic acids are stable molecules which enable living organisms to reproduce their entire genetic code from one generation to the next. They also allow for all subsequent proteins necessary for survival to be made as required.

(a) Differences between DNA and RNA

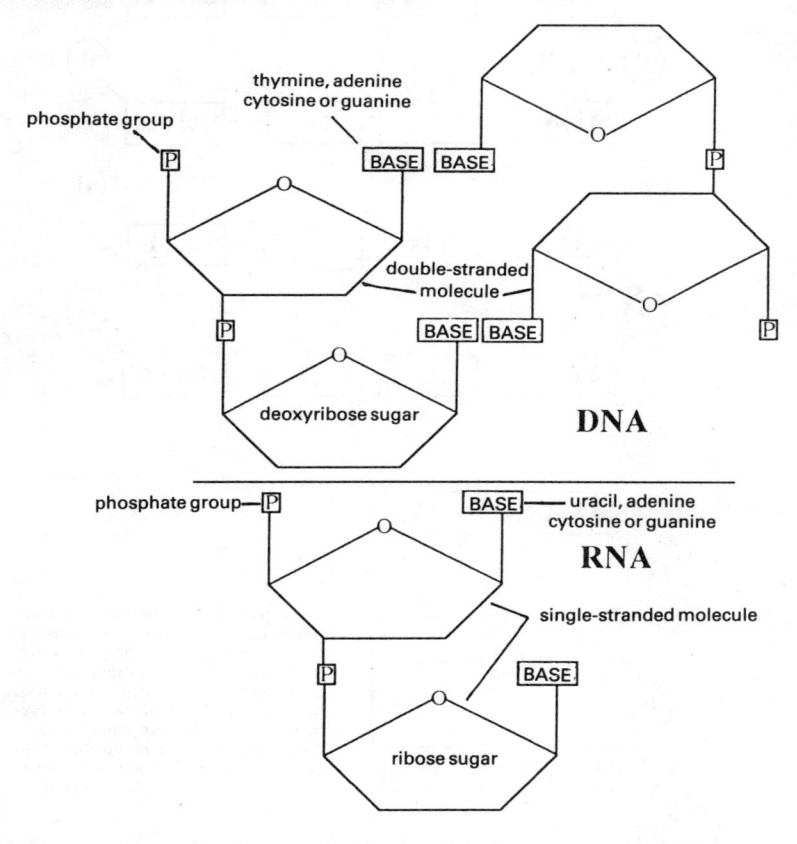

DNA exists as one type of molecule which is relatively large while RNA can exist as several different types of molecule such as mRNA and tRNA which are relatively small.

(b) Structure of DNA in relation to replication

The primary structure of DNA consists of the sequence of nucleotides held together to form a double helix in which two helices are wound round each other. The basic structure is shown below:

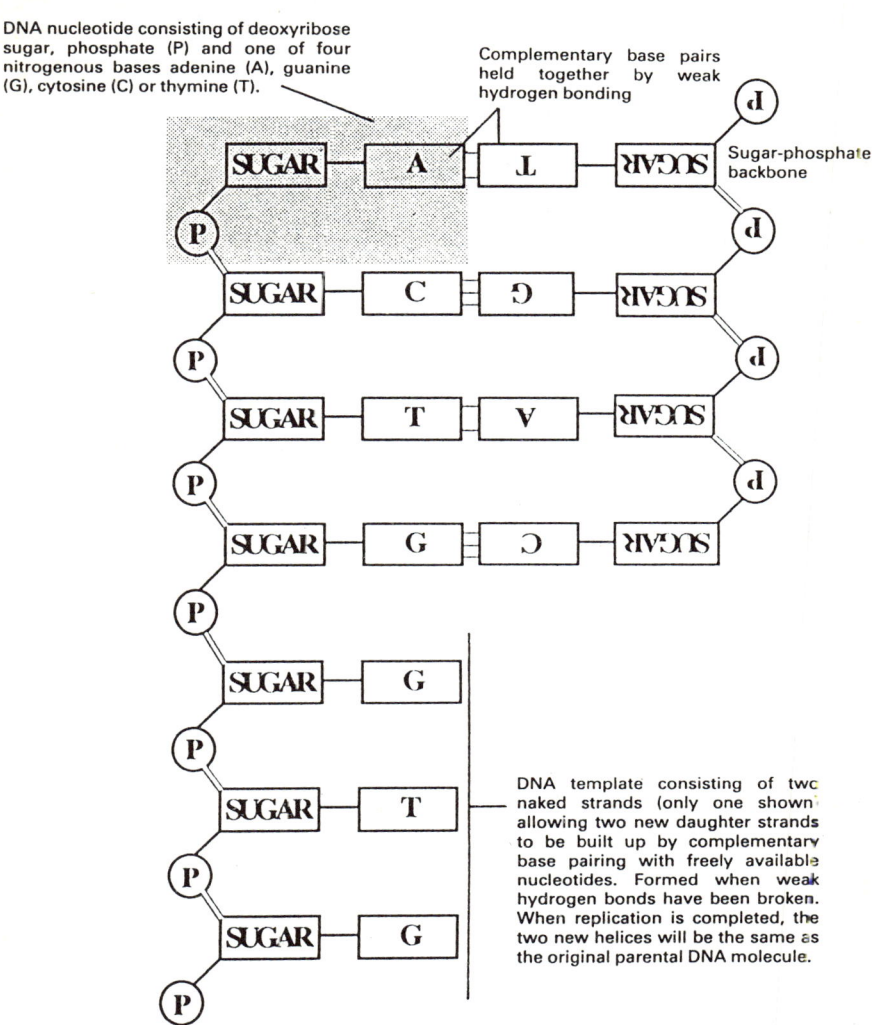

DNA nucleotide consisting of deoxyribose sugar, phosphate (P) and one of four nitrogenous bases adenine (A), guanine (G), cytosine (C) or thymine (T).

Complementary base pairs held together by weak hydrogen bonding

Sugar-phosphate backbone

DNA template consisting of two naked strands (only one shown) allowing two new daughter strands to be built up by complementary base pairing with freely available nucleotides. Formed when weak hydrogen bonds have been broken. When replication is completed, the two new helices will be the same as the original parental DNA molecule.

The basis of the genetic code is the linear sequence of the bases which has enough redundancy to code for all the naturally occurring amino acids.

2. GIVE AN ACCOUNT OF THE TRANSPORT SYSTEM UNDER THE FOLLOWING HEADINGS:

(a) need for a transport system; (7)
(b) basic organisation; (4)
(c) formation of tissue fluid. (4)

Introduction

Transport in humans is carried out by the circulatory system. This consists of the heart which forces blood to move inside blood vessels carrying useful materials to, and waste materials from, all body cells. The circulatory system also helps regulate and protect body function.

RELEVANT POINTS

(a) Need for a transport system

Human cells are highly specialised for particular roles,
    Such as transport, contraction and nerve conduction,
    Making them less able to carry out an independent existence,
    Which requires the ability to obtain food, avoid extreme changes in pH or temperature.

Body surface area in relation to volume is very small,
    With most cells being located deep within body tissues,
    And therefore unable to rely on diffusion to exchange nutrients, wastes or respiratory gases.

Transport system therefore needed to,
    Transport materials to and from cells,
    To allow them to carry out their basic metabolic processes,
    Control cell function via hormones and regulatory molecules,
    Maintain optimum environmental temperature and pH for cell function,
    Protect cells from damage by toxic chemicals or pathogens.

(b) Basic organisation

Transport system consists of the heart,
- Arteries and arterioles which carry blood away from the heart,
- Veins and venules which carry blood towards the heart,
- Capillaries which connect arterioles and venules,
- Lymphatic vessels which return tissue fluid back to the general circulation.

(c) Formation of tissue fluid

Large amounts of tissue fluid are constantly being produced from the blood,
- And reabsorbed back into the blood,
- Or the lymphatic system.

Tissue fluid is formed when certain components of blood plasma filter out from the arterial end of the capillary bed into the space outside,
- To bathe cells and supply them with nutrients.

This filtration results from the pressure of the blood within the capillaries,
- Which forces fluid across the permeable walls of the capillaries into the extracellular spaces.

SPECIMEN ANSWERS — QUESTION 2

EXAMPLE 1

Transport in humans is carried out by the circulatory system. This consists of the heart which forces blood to move inside blood vessels carrying useful materials to, and waste materials from, all body cells. The circulatory system also helps regulate and protect body function.

(a) Need for a transport system

Human cells are highly specialised for particular roles, such as transport, contraction and nerve conduction. This specialisation makes cells less able to carry out an independent existence which would require for example the ability to obtain food or avoid extreme changes in pH or temperature.

The surface area of the body in relation to its volume is very small with most cells located deep within body tissues. These cells are therefore unable to rely on diffusion to exchange nutrients, wastes and respiratory gases.

In order to transport materials to and from cells, allow them to carry out their basic metabolic processes and control cell function via hormones and regulatory molecules, a transport system is needed. The transport system also helps maintain optimum environmental temperature and pH for cell function as well as protecting cells from damage by toxic chemicals or pathogens.

(b) Basic organisation

The transport system consists of the heart, arteries and arterioles which carry blood away from the heart, venules and veins which carry blood towards the heart and capillaries which connect arterioles and venules. Lymphatic vessels return tissue fluid back to the general circulation.

(c) Formation of tissue fluid

Large amounts of tissue fluid are constantly being produced from the blood and reabsorbed back into the blood or lymphatic system. Tissue fluid is formed when certain components of blood plasma filter out from the arterial end of the capillary bed into the space outside. This tissue fluid bathes cells and supplies them with nutrients. Filtration results from the pressure of the blood within the capillaries forcing fluid across the permeable walls of the capillaries into the extracellular spaces.

EXAMPLE 2 (with a diagram)

Transport in humans is carried out by the circulatory system. This consists of the heart which forces blood to move inside blood vessels carrying useful materials to, and waste materials from, all body cells. The circulatory system also helps regulate and protect body function.

(a) Need for a transport system

Human cells are highly specialised for particular roles, such as transport, contraction and nerve conduction. This specialisation makes cells less able to carry out an independent existence which would require for example the ability to obtain food or avoid extreme changes in pH or temperature.

The surface area of the body in relation to its volume is very small with most cells located deep within body tissues. These cells are therefore unable to rely on diffusion to exchange nutrients, wastes and respiratory gases.

In order to transport materials to and from cells, allow them to carry out their basic metabolic processes and control cell function via hormones and regulatory molecules, a transport system is needed. The transport system also helps maintain optimum environmental temperature and pH for cell function as well as protecting cells from damage by toxic chemicals or pathogens.

(b) Basic organisation

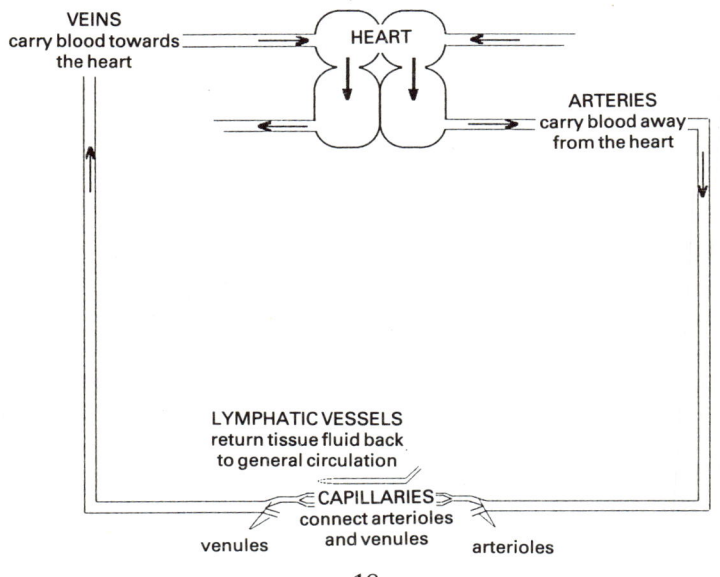

(c) Formation of tissue fluid

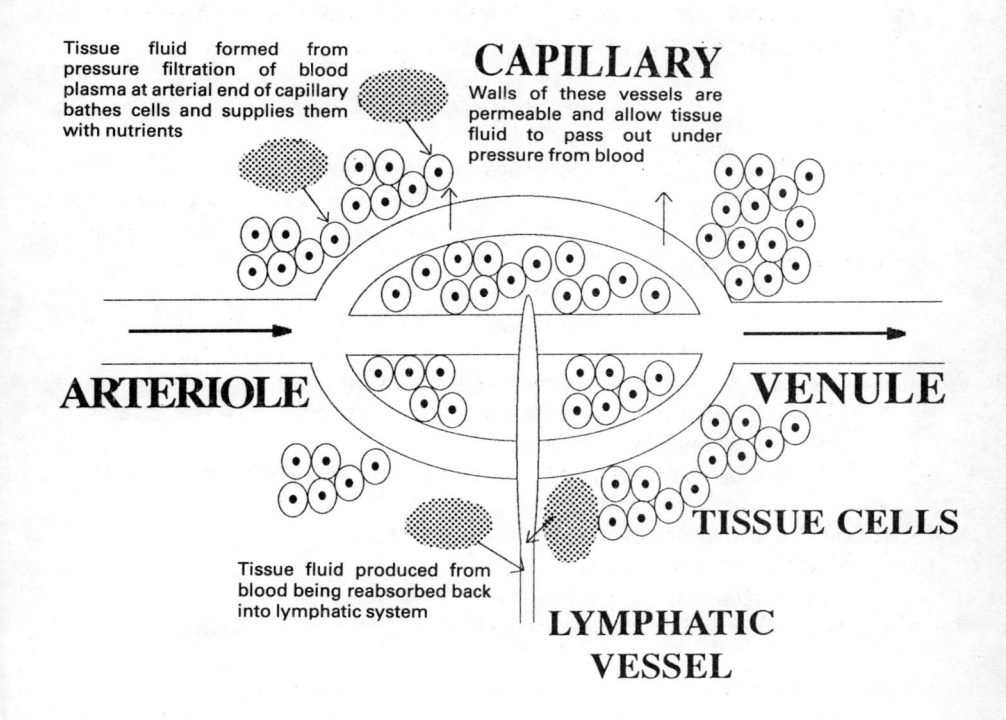

Tissue fluid formed from pressure filtration of blood plasma at arterial end of capillary bathes cells and supplies them with nutrients

CAPILLARY
Walls of these vessels are permeable and allow tissue fluid to pass out under pressure from blood

ARTERIOLE

VENULE

TISSUE CELLS

Tissue fluid produced from blood being reabsorbed back into lymphatic system

LYMPHATIC VESSEL

3 DISCUSS THE ROLE OF ENZYMES AS CATALYSTS IN BIOCHEMICAL PATHWAYS
(15)

Introduction

The role which enzymes play in biochemical processes is characterised by the variety and abundance of these proteins in living systems. They control every reaction necessary for survival and their study allows the mapping of metabolic pathways.

RELEVANT POINTS

Catalysts serve to increase unfavourable reaction rates,
    Allowing reactions to proceed only in the presence of the catalyst,
    And are essential in living systems,
    Where most pathways consist of catabolic and anabolic reactions.

Catalysts have distinct properties such as being,
    Effective in small amounts,
    And unchanged by the reaction.
Catalysts do not affect the position of equilibrium in a reaction.

Enzymes as catalysts have further properties such as being,
    Usually very specific both in the nature of the reactions being catalysed,
    And the structure of the substrate,
    And able to function under moderate conditions,
    But denatured by extremes of pH and high temperatures.

Specificity is shown by the "lock-and-key" mechanism of enzyme action,
    The substrate fitting the active sites,
    Which occupy only a small fraction of the surface of enzymes.
The enzyme-substrate complex dissociates to form the enzyme and products.
The enzyme is now free to enact further catalysis.

Enzymes determine the order of chemical reactions within the cell,
> Which also ensures that reactions take place in the correct locations and compartments,
> So that essential respiratory pathways,
> Such as the Krebs' cycle,
> Take place in the mitochondrial matrix,
> While the electron transport chain,
> Takes place on the inner mitochondrial membrane.

Enzymes have a regulatory function in biochemical pathways,
> Via feedback mechanisms,
> So that the activity of "early" enzymes in a pathway,
> May be controlled by later products,
> Ensuring that the products of reactions,
> Are present only when required,
> Which requires that enzymes work in sequence,
> The product of one enzyme-controlled reaction,
> Serving as a substrate for the next.

In order for all metabolic processes to operate together,
> For maximum efficiency of the body,
> Enzyme action must be under the influence of external factors,
> Such as oxygen concentration in respiring tissues.

4 DESCRIBE THE FACTORS WHICH DETERMINE ENZYME ACTIVITY UNDER THE FOLLOWING HEADINGS:
(a) inhibition; (8)
(b) concentration of enzyme and substrate. (7)

Introduction

Enzyme activity is affected by factors which change the molecular conformation of the enzyme. Such factors include the presence of inhibitors, heat and changes in pH. The relative amounts of enzyme, substrate and inhibitor present affects the rate of an enzyme reaction.

RELEVANT POINTS

(a) Inhibition

Most enzymes can be inhibited,
  Their activity being decreased by certain chemical agents called inhibitors.
Inhibitors can be reversible or irreversible,
  With reversible inhibition being either competitive or non-competitive.

In competitive inhibition the inhibitor competes,
  With the normal substrate for the enzyme,
  Is structurally similar to the normal substrate,
  Fits into the active site,
  And combines with the enzyme,
  But is not similar enough to substitute fully,
  For the normal substrate to form reaction products.
Competitive inhibition can be reversed by increasing the substrate concentration.

In non-competitive inhibition the inhibitor binds,
  At a site other than the active site,
  Rendering the enzyme inactive,
  By altering its shape.

Reversible inhibition is temporary,
    And does not permanently damage the enzyme.
Irreversible inhibitors,
    Combine chemically with a functional group on the enzyme,
    Permanently inactivating or destroying it.
Many poisons are irreversible inhibitors.

(b) Concentration of enzyme and substrate

Concentration of an enzyme is determined by,
    Measuring the rate of an enzyme reaction,
    Which is proportional to the concentration of enzyme,
    Providing excess substrate is present.
If enzyme concentration is constant,
    The initial rate of reaction is proportional to the substrate concentration,
    Up to a limiting value.

The total amount of enzyme in a cell at any given time is regulated by,
    The rate at which the enzyme is produced,
    And broken down in the cell.
Like all intracellular proteins,
    Enzyme molecules are broken down by proteolytic enzymes.

When the rate of enzyme production exceeds that of breakdown,
    Enzyme concentration is increased.

5. FUNCTIONAL ENZYMES ARE ESSENTIAL IN NORMAL METABOLISM. DISCUSS THIS STATEMENT UNDER THE FOLLOWING HEADINGS:
(a) activation of enzymes; (8)
(b) absence of particular enzymes. (7)

Introduction

Fine control exists for the functioning of enzymes in metabolism. This is essential for reactions to proceed in a controlled manner ensuring maximum energy efficiency in biochemical systems. The functions of particular enzymes can best be studied in cases where the enzyme is non-functional or absent.

RELEVANT POINTS

(a) Activation of enzymes

Some enzymes are only fully activated,
    By the presence of mineral ions, vitamins or other enzymes.

Mineral ions called co-factors,
    Are required for catalytic activity.
Co-factors may bind loosely,
    And reversibly with substrate,
    Or bind tightly and permanently,
    To the active site.
Metal atoms,
    Such as zinc, iron, copper or calcium,
    Can act as co-factors.

If the co-factor is an organic molecule,
    It is called a co-enzyme.
Most vitamins are co-enzymes,
    Or raw materials from which co-enzymes are made.
Vitamin K for example,
    Is essential for full functioning of the blood-clotting mechanism.

Most enzymes are activated by hydrolysis of inactive precursors,
    Called pro-enzymes.
Digestive enzymes,
    Such as chymotrypsin,
    Have this regulatory mechanism.

The blood-clotting cascade,
> Is a series of pro-enzyme activations,
> Each activated enzyme activating the next in the cascade,
> Allowing for the amplification of each step,
> And the operation of fine control of the blood-clotting process.

(b)   Absence of particular enzymes

Biochemical systems involve multi-enzyme reactions which allow,
> Fine control of processes,
> And the modification and possible removal,
> Of toxic intermediates.

All enzymes involved in multi-enzyme reactions,
> Are necessary for the full,
> Completion of the reactions,
> And function of feedback mechanisms.

Non-functional or absent enzymes,
> Can lead to inborn errors of metabolism,
> Such as phenylketonuria,
> Which is recessively inherited,
> Affecting 1 in 15,000 births.

The non-functional enzyme leads to the inability to break down the amino acid phenylalanine,
> As part of the normal metabolism of protein.

Phenylalanine and the by-product phenylpyruvic acid,
> Can accumulate to toxic levels in the blood,
> Causing severe mental retardation.

All newborns are now tested,
> And if phenylketonuria is detected,
> They are treated with a special diet,
> Lacking phenylalanine.

In contrast, failure of a different enzyme in the same reaction,
> Leads to the benign condition,
> Called alcaptonuria,
> Which was the first described genetic disease.

Alcaptonuria results in the presence of a black melanin substance,
> In the urine.

6 PROTEINS HAVE MANY DIVERSE ROLES IN LIVING SYSTEMS. DISCUSS THIS STATEMENT UNDER THE FOLLOWING HEADINGS:
   (a)   intracellular proteins; (7)
   (b)   proteins outwith the cell. (8)

Introduction

The variety and distinctive characteristics of proteins are central to the functioning and maintenance of all biological systems. Although proteins have very varied roles, they are commonly made from the same twenty amino acids.

RELEVANT POINTS

(a)   Intracellular proteins

Almost all enzymes are proteins.
Enzymes catalyse chemical reactions,
   Such as those of glycolysis,
   The Krebs' cycle,
   And the digestive functions of the lysosome.

Contractile proteins,
   Such as actin and myosin,
   Are involved in co-ordinated motion of muscle fibres.

Nucleoproteins bind to nucleic acid
   And are involved in replicative functions.

Structural proteins,
   Link the plasma membrane to the sub-membrane network of actin filaments.

The protein tubulin,
   Forms the basis of microtubule assembly of spindle fibres,
   During cell division.

Some proteins allow intracellular storage of ions,
   Such as calcium and iron,
   Which is stored in the liver,
   As a complex with a protein,
   Called ferritin.

(b) Proteins outwith the cell

Many proteins function as extracellular enzymes,
    For example the digestive enzymes,
        Such as pepsin and trypsin.

The transport of substances,
    Such as glucose, fatty acids, amino acids, ions, toxins,
    Employs proteins.
    For example, the main plasma protein,
    Albumen.
The transport of steroids and thyroid hormones,
    As well as drugs,
    Such as penicillin and aspirin,
    Is mediated by proteins.

Proteins are essential in osmoregulation,
    Of extracellular fluid.

Proteins aid the transport to the liver,
    Of insoluble fatty acids,
    And bilirubin,
    A breakdown product of haemoglobin.

Structural proteins,
    Such as collagen and elastin,
    Form an integral part of cells and tissues,
    While keratin,
    Is found in the skin, liver and nails.

Proteins are also involved in defence mechanisms,
    For example as antibodies,
    Which protect the body against disease,
    And fibrinogen in plasma,
    Which is important in blood clotting.

Some hormones such as insulin and growth hormone,
    Are proteins,
    As are the cell receptors for these hormones.

The response of nerve cells to specific stimuli,
    Is mediated by receptor proteins.

7 DISCUSS THE IMPORTANCE OF ENERGY BEING RELEASED IN A CONTROLLED WAY
(15)

Introduction

The controlled release of energy is essential in living systems for maintenance and survival. The processes of metabolism must interact in such a way as to benefit the whole organism so that energy is conserved and utilised as efficiently as possible.

RELEVANT POINTS

Energy is required for the metabolism of all cells,
    Which must be able to grow,
    Maintain and repair themselves,
    Reproduce and respond to stimuli.

Energy must be available for cellular work,
    Mechanical work of muscular contraction,
    Electrochemical work of conducting nerve impulses,
    Active transport to move molecules against a concentration gradient,
    Chemical work of synthesising new molecules for growth,
    Or for energy storage.

Random energy release,
    Or the sudden release of large amounts of heat,
    Would be inappropriate to the order essential to the human body,
    Which is homeothermic.
Cells contain effective control mechanisms,
    Which minimise the energy lost as heat,
    By storing energy in chemical bonds,
    Of complex organic molecules,
    Such as adenosine triphosphate.

Since metabolism consists of both catabolic and anabolic reactions,
    Each must be controlled in such a way that couples both reactions.

Adenosine triphosphate is the common currency,
    Which allows favourable chemical reactions,
    Such as cellular respiration,
    To be coupled with unfavourable anabolic processes,
    Such as biosynthesis.

Most chemical reactions are reversible,
    But a variety of control mechanisms in a cell,
    Ensure that energy is available,
    For a particular reaction to proceed in the direction required,
    And at the correct time.

8. THE BODY SOMETIMES UTILISES COMPOUNDS OTHER THAN GLUCOSE AS SOURCES OF ENERGY. DISCUSS THIS STATEMENT UNDER THE FOLLOWING HEADINGS:
   (a) marathon running; (7)
   (b) starvation. (8)

Introduction

Although glucose from carbohydrate intake is the main source of energy, when energy output is greater than intake, the body must draw upon its stored materials to maintain its metabolic rate and survive.

RELEVANT POINTS

(a) Marathon running

Presence of adrenaline,
    Helps liberate glucose from glycogen stores,
    To be used for muscle action,
    But are quickly exhausted.

Excessive muscle respiration will produce lactic acid,
    Which will be converted to pyruvic acid,
    In the liver,
    And used to provide energy.

The body next adapts to fat metabolism,
    With oxidation of fatty acids,
    Triglycerides in adipose tissue,
    Broken down to fatty acids,
    Which are then bound to albumen,
    Transported to the liver,
    Converted to acetyl-CoA,
    Used to provide energy in the liver,
    And other cells via the Krebs' cycle.

(b) Starvation

Three stages are involved in starvation,
    Initial, adaptation and terminal.

Initially the body uses up its glycogen stores,
> Adapting then to the oxidation of its fat stores,
> Using long term fatty acid oxidation,
> Which depletes the body of fat soluble vitamins,
> Such as vitamin A,
> Essential for vision,
> Vitamin D,
> Essential for bone formation,
> And vitamin K,
> Essential for efficient blood clotting.

Terminally the body utilises body protein,
> Which then leads to protein energy malnutrition.

Protein is used for energy only when other stores are exhausted,
> And is broken down to its constitutive amino acids,
> Which are transaminated in the liver,
> And used to provide energy,
> Via the Krebs' cycle.

Protein breakdown and inadequate protein intake,
> Lead to lack of essential amino acids,
> Which means that the liver cannot make plasma proteins,
> Leading in turn to inefficient antibody production,
> And osmotic imbalance,
> Causing fluid accumulation in tissues,
> With possible mental and physical retardation.

When body proteins have been depleted to around half the normal level,
> Death occurs,
> Usually from failure of the heart muscle.

9

CELL MEMBRANES ARE FLUID AND POROUS IN NATURE. DISCUSS WHY THIS IS IMPORTANT IN RELATION TO THE FUNCTION OF EACH OF THE FOLLOWING:
- (a) plasma membrane; (6)
- (b) rough endoplasmic reticulum; (5)
- (c) Golgi apparatus. (4)

Introduction

The membranes of cells and internal organelles are responsible for the organisation and separation of all cellular biochemical functions. They also ensure that the cell composition is that required for the cell's own activities.

RELEVANT POINTS

(a) Plasma membrane

Plasma membrane must be porous and fluid,
    To function as a selective barrier,
    Regulating chemical composition,
    And pH of the cell,
    By allowing some substances to pass readily,
    Between cell and external environment,
    While impeding the entrance and exit of others.
The membrane must allow sufficient traffic,
    Of oxygen, nutrients and waste products.
It must also flex, stretch, self-repair, grow, bud vesicles,
    Interact with cytoplasmic structures,
    And be capable of division.
The membrane must also orientate cell junctions,
    For communication between cells.

(b) Rough endoplasmic reticulum

Fluidity is essential to provide a large surface area inside the cell,
    For enzymatic activities,
    And to divide cell into compartments,
    For different biochemical functions.
The organelle has a granular appearance,
    Due to ribosomes being present on the outer membrane.
Ribosomes are the sites of protein synthesis.

Lipids are also synthesised in this organelle,
    And these go to build all other cellular membranes.
The rough endoplasmic reticulum must therefore bud vesicles,
    Which contain assembled proteins and lipids.
The fluid nature allows the transport of various chemical substances,
    From one part of the cell to another,
    Including regulating cytoplasmic calcium.

(c) Golgi apparatus

Fluidity is the key feature of the Golgi apparatus,
    Which functions mainly in processing and packaging,
    Protein compounds made in the rough endoplasmic reticulum.
This organelle is highly developed in secretory cells,
    Vesicles containing substances for secretion,
    Pass to the Golgi apparatus,
    To form new Golgi apparatus membranes,
    Which bud secretory vesicles,
    Moving to the plasma membrane,
    Or another membrane within the cell,
    Depending on the "address" given during processing.

10 DESCRIBE THE FUNCTIONS OF LIPIDS AND PROTEINS WITHIN THE PLASMA MEMBRANE
(15)

Introduction

The plasma membrane can be represented by the "fluid-mosaic" model visualised as proteins floating in a sea of lipid. Together these two components of all plasma membranes are responsible for the interactions necessary between the cell and the surrounding environment.

RELEVANT POINTS

Lipids confer fluidity to the cell membrane,
    And make up some 70% of its composition.

Three types of lipids exist,
    Phospholipid,
    Cholesterol,
    Glycolipid.
All have polar head groups and non-polar tail groups,
    And assemble tail to tail,
    Giving the bilayer structure,
    Characteristic of all membranes.
Membranes are held together by hydrophobic interactions between lipids.

Lipids diffuse in the plane of the membrane,
    Giving surface viscosity equivalent to that of light olive oil.
The temperature at which the membrane can solidify,
    Depends on the lipid composition,
    And the properties of the unsaturated phospholipids.
Cholesterol enhances fluidity.
Membranes must be at the correct viscosity,
    For efficient enzyme activation,
    And vesicle formation.

Lipids are involved in passive transport of substances.
Hydrophobic molecules,
    Such as hydrocarbons and oxygen,
    Can dissolve and pass through the lipids of the membrane easily.

Lipids are impermeable to all ions.

Proteins carry out all other plasma membrane functions.

There are two types of membrane proteins,
    Integral and peripheral proteins.
Integral proteins span the membrane,
    While peripheral proteins are attached to the surface of the membrane,
    And are involved in anchoring the membrane to the sub-membrane cytoplasmic structures.
A single integral protein may perform a combination of tasks such as,
    Active transport of substances to regulate pH and ion composition,
    Enzyme functions,
    Receptor sites for hormones,
    Cell adhesion forming intercellular junctions.

11. THE NEED FOR A CONSTANT ENVIRONMENT WITHIN THE CELL IS ESSENTIAL IN LIVING SYSTEMS. DISCUSS THIS STATEMENT

(15)

Introduction

As with the body as a whole, individual cells must maintain a homeostatic environment for their own metabolic processes. This involves careful regulation of cell composition.

RELEVANT POINTS

Optimum cellular temperature and pH,
    Must be maintained for effective enzyme action.
Co-ordinated metabolic action,
    And controlled heat dissipation,
    Ensure optimum cellular temperature.
pH is kept constant by selective ion uptake,
    By the plasma membrane.

Most of the volume of a cell is made up of water,
    Which is a solvent in chemical reactions.
Water can diffuse freely in and out of a cell,
    By the process of osmosis.
Normally cells are isotonic with their environment,
    Of blood plasma and body fluids.
This means that intracellular and extracellular osmotic pressures are the same,
    Resulting in no net movement of water molecules,
    Across the plasma membrane.
The correct water concentration of a cell,
    Is essential to provide the optimum concentrations of reactants,
    Involved in cellular processes.

In order for tissues to operate effectively,
    As found in co-ordinated muscle action for example,
    Cells within the tissue must maintain the correct metabolic rate,
    By balanced interaction with neighbouring cells.

Such cell-to-cell communication,
>   Ensures minimum fluctuation of cell composition,
>   And therefore no gross changes,
>   In individual cell activity.

Membrane-bound organelles partition the cytoplasm,
>   Acting as barriers,
>   To make it possible for the chemical environment in the organelle,
>   To differ from that in the general cytoplasm,
>   Which permits metabolic processes to proceed,
>   In an orderly effective manner.

12 USING AN EXAMPLE YOU HAVE STUDIED, WRITE AN ESSAY ON RECOGNITION OF ANTIGENS
(15)

Introduction

Self and non-self recognition is the basis for the body's action against all foreign substances in the processes of cellular responses in defence mechanisms. Recognition of antigens can best be illustrated by the ABO blood group system.

RELEVANT POINTS

Blood group is an important characteristic of a person's chemical identity.
Individuals of group A,
    Have the A antigenic molecule,
    On the surface of their red blood cells.
Group B,
    Have the B antigen.
Group AB,
    Have both A and B antigens.
Group O,
    Have neither A nor B antigens.
These surface molecules are antigenic only in the sense that they are foreign,
    If transferred to another individual.

A person with group B does not produce antibodies against the B antigen,
    But does produce A antibodies.
Thus it is that a group B individual receiving a transfusion,
    Containing A antigens,
    Would cause the donated blood to agglutinate,
    Via A antibodies,
    Forming clots,
    And endangering the recipient's life.

AB individuals can receive blood from all donors,
    Since they carry both antigens on their own red blood cells,
    And do not produce A or B antibodies.
AB individuals are therefore termed universal recipients.

O individuals can only receive blood from group O donors,
>   Since they produce both A and B antibodies.
These individuals are termed universal donors,
>   Because their blood cells carry neither antigen,
>   And cannot cause adverse reactions in recipients,
>   Even if they produce A and/or B antibodies.

Recognition of another blood group antigen called the Rhesus factor,
>   Can be important in pregnant women.
Problems develop when the mother lacks the Rhesus factor,
>   On her red blood cells,
>   When she is termed Rhesus-negative,
>   While her foetus does have the Rhesus factor,
>   Which is termed Rhesus-positive.
The mother develops antibodies against the Rhesus factor,
>   When foetal blood comes into contact with maternal blood.
The first such exposure produces a mild response in the mother,
>   But this response becomes more severe in subsequent pregnancies,
>   Due to the immunological memory of the mother.
Her antibodies can pass across the placenta,
>   Late in gestation,
>   And destroy the red blood cells of the foetus.

13 DISTINGUISH BETWEEN CELLULAR AND HUMORAL RESPONSES AS DEFENCE MECHANISMS

(15)

Introduction

The two main types of specific immunity involve responses by whole cells. These are known as humoral and cell-mediated immunity.

RELEVANT POINTS

The cells responsible for immune responses,
    Are white blood cells,
    Called lymphocytes.
There are two distinct populations of lymphocytes,
    B lymphocytes,
    And T lymphocytes.

As B and T lymphocytes mature in the bone marrow and thymus respectively,
    They develop immunocompetence,
    Each cell becoming competent at recognising a specific antigen,
    And making an appropriate immune response,
    If ever that antigen is encountered.

Humoral immunity involves immunocompetent B lymphocytes,
    Which are activated by specific antigens,
    To release soluble proteins,
    Called antibodies,
    Which bind to the antigen,
    To form an antigen-antibody complex.
The antigen-antibody complex is recognised,
    And destroyed by a variety of effector mechanisms,
    Such as phagocytosis.
This is called the primary response,
    And results in the cloning of activated B lymphocytes,
    Producing more antibody secreting cells,
    And also memory cells,
    Which persist after the first encounter with the antigen,
    To mount a secondary response,
    If that same antigen is encountered again.

The secondary response is faster and greater in magnitude.

The humoral system defends primarily against,
    Free bacteria and viruses,
    Present in body fluids.

Cell-mediated immunity is carried out by highly specialised cells,
    Called T lymphocytes,
    Which circulate in the blood,
    And are present in lymphoid tissue.
The main function of T lymphocytes is to destroy cells,
    Already infected with a particular antigen.
T lymphocytes cannot be activated by free antigens.
Immunocompetent T lymphocytes can respond,
    Only to bacteria and viruses which have already infected host cells,
    And also fungi and protozoa.

14 DESCRIBE THE FUNCTION OF MACROPHAGES (15)

Introduction

Macrophages are important in the "front line" defence of the body. They are activated by inflammation either locally or involving the whole body, functioning as non-specific "cleaners" of the interstitial fluid.

RELEVANT POINTS

Macrophages are amoeboid cells,
    Which travel through extracellular fluid,
    Engulfing bacteria, viruses and cell debris,
    By the process of phagocytosis.
These cells trap foreign particles,
    By the use of multiple long projections,
    Called pseudopodia,
    Ingesting the particle by enclosing it in a membrane,
    Pinched off from the plasma membrane,
    A process known as endocytosis.
Several lysosomes,
    Adhere to the membrane of the vesicle,
    And eventually fuse with it.
The lysosomes release their powerful digestive enzymes,
    Onto the invading particle,
    While the vesicular membrane releases hydrogen peroxide.
These substances destroy the particle,
    Breaking down its macromolecules,
    To smaller harmless molecules,
    Which can then be released,
    Or used by the macrophage in anabolic processes.

A macrophage can phagocytose about one hundred bacteria during its lifespan,
    Eventually becoming inactivated,
    By leaking lysosomal enzymes.

Some bacteria can counteract the action of macrophages,
    By releasing enzymes which destroy the lysosomal membranes.

The causative bacterium of tuberculosis,
    Possesses a cell wall which resists the action of lysosomal enzymes.

Some macrophages are constantly motile,
    Releasing activated agents when appropriate.
Others stay in one place,
    Destroying bacteria which pass by.

The air-sacs in the lungs,
    Contain large numbers of non-motile macrophages,
    Which destroy foreign particles which enter with inhaled air.

15 DESCRIBE THE TWO FORMS OF ACQUIRED IMMUNITY (15)

Introduction

Active immunity can be acquired naturally or artificially. Both ways involve an initial immune response to a pathogen and the subsequent development of memory cells to give long-lasting immunological protection against that specific pathogen.

RELEVANT POINTS

In naturally acquired immunity,
    Pathogens enter the body,
    Through a natural encounter,
    By a variety of transmission routes.
These routes include contact,
    Which needs an abrason,
    Inhalation of aerosol droplets,
    Such as coughs and sneezes,
    Ingestion,
    Inoculation,
    By the bite of an animal.
The initial infection can give rise to the full symptoms of the disease,
    And a full immune response,
    Such as fever,
    And the characteristics of inflammation,
    Which include redness, heat, oedema and pain.

Artificially induced immunity,
    Is the basis of immunisation,
    Which utilises vaccination.
Several types of vaccination exist,
    For example killed or live.
In the killed vaccination,
    The pathogen is killed often by the use of a chemical,
    Such as formaldehyde,
    But still retains its surface antigens,
    And is therefore capable of eliciting an immunological response.

Killed vaccines are used against whooping cough and typhoid fever.

Killed vaccines are less effective than live vaccines,
> Which utilise an attenuated (weakened) pathogen.

Pathogens may be attenuated by passing through cells of a non-human host,
> When mutations occur which adapt the pathogen to the non-human host,
> > So that it can no longer cause the full disease in humans.

Live vaccines give real infections,
> But with reduced symptoms,
> > In diseases such as polio, smallpox and measles.

Modified toxins called toxoids,
> May be used to immunise against tetanus and diphtheria.

16. DISCUSS IMMUNITY UNDER THE FOLLOWING HEADLINES:
(a) autoimmunity; (6)
(b) allergy. (9)

Introduction

Sometimes the immune system does not operate as required for the good of the body, turning against "self" in autoimmune diseases and conferring hypersensitisation to environmental antigens.

RELEVANT POINTS

(a) Autoimmunity

In autoimmune diseases immune reactions develop,
    Against components of body cells.
Rheumatoid arthritis may occur,
    Due to an immune response against the cells in the joints,
    Resulting in inflammation.
Juvenile diabetes,
    Is caused by an immune response against the β-cells of the pancreas.
Viruses, drugs or genetic mutations,
    May alter the surface antigens of some cells,
    Enough to cause them to be recognised as foreign.
Another cause of autoimmunity may be the unmasking of cells,
    From tissues not normally in contact with the immune system,
    And therefore foreign,
    Such as injury to the cornea,
    Which exposes certain molecules,
    Resulting in the cornea becoming opaque.

(b) Allergy

Allergies are hypersentitivities of the body's defence system,
    To certain antigens called allergens.
Allergic reactions are very rapid,
    And sensitive to even minute quantities of the allergen.

Typical allergens are pollen or animal dander.
Reactions can occur in the nasal and bronchial passages,
    As well as the gastrointestinal tract and skin.
Antibodies of the IgE family participate in allergic reactions,
    Binding to the surface of mast cells,
    Found in connective tissue.
When the antibody binds to the allergen,
    Degranulation of the mast cell occurs,
    And it releases histamine,
    An inflammatory substance,
    Which causes smooth muscle to contract,
    And blood vessels to dilate and become leaky.
Histamine is also responsible for the symptoms of sneezing,
    Itchiness of the skin,
    And watering of the eyes.
Antihistamine drugs interfere with this action.
Acute allergic responses can cause anaphylaxis,
    For example in the case of bee-stings,
    Which is life-threatening,
    Because it involves sudden degranulation,
    And abrupt dilation of blood vessels,
    Which causes rapid blood-pressure drop,
    Leading to potentially fatal shock reactions.
Anaphylaxis may be countered by adrenaline injection.

17 WHAT IS THE NATURE OF VIRUSES AND HOW DO THEY INVADE CELLS?

(15)

Introduction

Viruses are intracellular parasites which, by a variety of mechanisms, can utilise the host cell's "machinery" to replicate themselves thereby producing many more viruses to the detriment of the host.

RELEVANT POINTS

A virus can exist in two different forms,
    A static phase,
    And a dynamic phase.
In the static phase the virus particle (virion) has no active functions,
    And comprises the nucleic acid genome,
    Surrounded by a protein coat,
    Called the capsid,
    Some having an outer membranous envelope.
The capsid coat serves to protect the genome.

The dynamic phase consists of the infected cell,
    In which the virion is dismantled,
    Allowing transcription and replication of the genome,
    Leading to synthesis and release of new virions.

Viruses are characterised by,
    Their small size,
    The largest (smallpox) at 300nm,
    While the majority are less than 150nm,
    And the smallest (parvovirus) at 20nm,
    And by the fact that they contain only one type of nucleic acid,
    Which can be DNA or RNA,
    But not both.

Viruses also have a simple structure.
The virion is metabolically inert,
    No energy generating system is present,
    Nor protein synthesising system.

Viruses are obligate parasites,
    And therefore can only replicate inside a host cell.
A viral infection of a host cell comprises,
    Attachment to the surface of the host cell,
    Via specific receptors,
    Giving rise to virus-host specificity,
    Followed by penetration of the host cell,
    Which in animal cells is by the process of phagocytosis.
After penetration,
    Replication,
    Using host cell ribosomes,
    Energy,
    And many of its enzymes takes place,
    Which gives rise to the production of newly synthesised viral components,
    To complete the production of new viruses.
Viral release is by lysis of the host cell membrane,
    By a viral enzyme,
    Which is made late in the replication process,
    And causes the death of the infected cell.

18. BY MEANS OF AN EXAMPLE YOU HAVE STUDIED, OUTLINE THE LIFE CYCLE OF A VIRUS AND ITS EFFECTS ON THE HOST CELL

(15)

Introduction

Understanding viruses, the mechanisms of their infection and resulting pathogenesis, allows anti-viral agents to be designed which may counteract the virus-host specificity common to all viruses. Research with retroviruses in particular is providing an insight into the mechanisms involved in all forms of cancer.

RELEVANT POINTS

One of the best studied viruses is the human immunodeficiency virus,
- Commonly known as HIV,
- Which leads to the condition called acquired immune deficiency syndrome (AIDS).

HIV is a retrovirus.

Replication cycle of such a retrovirus involves,
- Attachment to a host cell,
- Via a specific receptor,
- Entry by fusion of the viral envelope with the host cell membrane,
- Uncoating of the viral capsid in the host cytoplasm.

The viral enzyme uses viral RNA as a template,
- To make single strands of DNA,
- Which are then used as templates,
- To complete a DNA double helix,
- A process known as reverse transcription.

The viral DNA enters the nucleus of the host cell,
- And integrates into the chromosomal DNA of the host,
- Becoming a "provirus".

Proviral DNA is transcribed into viral RNA,
- And translated into viral proteins,
- With subsequent assembly of new capsids,
- And viral RNA with associated proteins.

Virions bud from the plasma membrane,
- To form new retroviruses.

The virus which causes AIDS infects and eventually kills,
   Helper T cells,
   Whose function is to activate other T and B lymphocytes,
   Responsible for cellular and humoral immunity to infection.

Persons with AIDS cannot effectively form antibodies,
   And may die usually from opportunistic infections,
   Or cancers,
   Which normally do not present a threat to a healthy immune system.

HIV also infects macrophages,
   Including those in the brain,
   Which may lead to neurological disorders.

The AIDS virus has a long incubation period,
   So that it may be integrated into the host cell genome,
   For months or even years.
Activation of T lymphocytes may activate the provirus,
   Leading to the assembly,
   And release of new virus particles.

19 COMPARE AND CONTRAST MITOSIS AND MEIOSIS (15)

Introduction

Many of the steps of meiosis closely resemble corresponding steps in mitosis. However, the chromosome number is reduced by half in meiosis but not in mitosis. This is the key difference in terms of genetic consequences.

RELEVANT POINTS

Meiosis and mitosis are similar in that both are preceded by replication of chromosomes,
> With the events unique to meiosis occurring,
> During the first division,
> Called meiosis I.

The mechanisms involved in the second meiotic division,
> Called meiosis II,
> Are essentially the same as the processes in mitosis.

The main differences between meiosis and mitosis may be summarised as follows:

Mitosis	Meiosis
One cell division.	Two cell divisions.
Number and types of chromosomes in daughter cells are the same as in parent cell.	Number of chromosomes in daughter cells is half that of parent cell and different in type.
The diploid number is maintained.	Haploid number results.
Homologous chromosomes do not pair up.	Homologous chromosomes pair up (synapsis).
No chiasmata form.	Chiasmata may occur.
No crossing-over takes place.	Crossing-over may take place.
There is no genetic exchange between chromatids.	There may be possible genetic exchange between non-sister chromatids.
Two daughter cells are formed.	Four daughter cells are formed.
Occurs in body cells.	Occurs in gamete forming tissue.

Mitosis	Meiosis
There is no variation between daughter cells.	Variation between daughter cells is possible.
Single line of chromosomes aligned on equator.	Double line of chromosomes aligned on the equator.
Single spindle formed during whole process.	Two spindles are formed during the whole process.
One replication of centrioles.	Two replications of centrioles.

20 DESCRIBE THE FACTORS LEADING TO VARIATION UNDER THE FOLLOWING HEADINGS:
    (a)   independent assortment;   (6)
    (b)   crossing-over;   (4)
    (c)   mutation.   (5)

Introduction

Variation within a population is important to allow the process of natural selection of individuals best suited to their environment. Genetic variation arises through sexual reproduction and mutation.

RELEVANT POINTS

(a) Independent assortment

Sexual reproduction allows for genetic variation.
At metaphase in meiosis I,
    Each homologous pair of chromosomes,
    Which consist of one maternal and one paternal chromosome,
    Pair up.
The orientation of the homologous pair relative to the two poles of the cell,
    Is random.
There are thus two possibilities,
    An equal chance that a particular daughter cell of meiosis I,
    Will receive either the maternal or paternal chromosome,
    Of a certain homologous pair.
The gametes produced may have any of the possible combinations of maternal and paternal chromosomes.
In humans meiosis independently assorts twenty-three pairs of chromosomes,
    So that the number of possible combinations of maternal and paternal chromosomes in gametes is 2^{23},
    Which represents around 8 million different combinations.

(b) Crossing-over

Genetic variation through sexual reproduction,
    Occurs during prophase of meiosis I.

Pairing (synapsis) is very precise,
>And results in genetic information being exchanged,
>Between adjacent chromatids,
>Of paired homologous chromosomes.
>Crossing-over is usually a reciprocal process,
>>In which identical lengths of chromatids,
>>Are exchanged,
>>Without loss of genetic information.

Crossing-over events occur at least once,
>In most human chromosomal pairings during meiosis,
>Leading to extensive variation in the gametes.

(c) Mutation

Mutation is an important process leading to genetic variation,
>And forms the basis of natural selection in the evolution of species.

A mutation which affects any gene locus,
>Is entirely random.

Most harmful mutations occur in somatic cells,
>Which usually die within an individual.

Most point mutations,
>Which affect a single base in the DNA,
>Are harmless,
>Due to redundancy in the genetic code.

Most of the human DNA is "silent",
>That is, does not code for protein synthesis.

Beneficial mutations are rare.

A mutation which affects a protein enough to alter its function,
>Is more likely to be harmful than beneficial.

There may be a link between certain mutations,
>And some cancers.

Mutations may be induced by mutagens,
>Such as ionising radiation and certain chemicals.

21 DISCUSS INHERITANCE UNDER THE FOLLOWING HEADINGS:
    (a)   patterns of dominance;   (8)
    (b)   multiple alleles.   (7)

Introduction

The mechanisms of inheritance of dominant and recessive alleles was proposed by Gregor Mendel. For some traits however, intermediate inheritance may be observed, where the path from genotype to phenotype has some complications.

RELEVANT POINTS

(a) Patterns of dominance

Patterns of dominance may be complete or incomplete.
Complete dominance occurs when two alleles of a pair are different,
    As found in the heterozygous condition,
    One, the dominant allele,
    Being fully expressed,
    While the other, recessive allele,
    Is completely masked.

In incomplete dominance a recessive allele may be expressed under certain conditions,
    As shown in the sickle-cell trait.
Heterozygotes for this condition,
    Express a normal phenotype,
    Under normal conditions of oxygen tension.
Sickling may occur in the red blood cells of heterozygotes,
    Under conditions of low oxygen tension,
    As is found in high altitudes.

Another example of an inherited disease,
    Called familial hypercholesterolaemia,
    Shows incomplete dominance.
Individuals homozygous for this condition,
    Have six times the normal blood cholesterol level,
    And die in infancy.

Heterozygotes for this trait,
> Have one normal allele,
> And one defective allele,
> Which results in them having twice the normal levels,
> Of blood cholesterol,
> And being prone to atherosclerosis.

(b) Multiple alleles

Some genes can exist in more than two allelic forms,
> As found in the ABO blood types.

There are four phenotypes for this trait,
> An individual's blood type may be A, B, AB or O.

The letters refer to gene products,
> Which are antigenic molecules,
>> Found on the surface of red blood cells.

The four blood types result from various combinations,
> Of three different alleles.

Both alleles encoding types A and B,
> Are dominant to the allele O,
> Which encodes neither A nor B.

Individuals heterozygous for alleles encoding A and O,
> Or encoding B and O,
> Have blood types A and B respectively.

People homozygous recessive for O,
> Will have blood type O,
> As gene products A and B are not present.

The alleles encoding for blood types A and B,
> Are said to be co-dominant,
> As both will be expressed in a heterozygote for these alleles,
> Resulting in blood type AB.

22 DESCRIBE MONOHYBRID INHERITANCE IN HUMANS (15)

Introduction

Monohybrid inheritance describes the pattern of inheritance of a pair of alleles where one allele is dominant and one allele is recessive.

RELEVANT POINTS

The progeny of the parental generation (P generation),
    Is called the first filial generation (F_1 generation).
The subsequent generation produced by F_1 matings,
    Are termed the F_2 generation.
The factors transmitted from parents to progeny,
    Through the gametes,
    Are termed genes.
For each gene there is one form (allele),
    For the paternal trait,
    And another allele for the maternal trait.

In a monohybrid cross,
    All the F_1 progeny are phenotypically the same,
    As one of the parents.
The expression of the "missing" parental trait,
    Is masked by the visible trait,
    A feature called dominance.
One parental allele is dominant,
    And the other is recessive.

In the F_2 generation,
    Both parental phenotypes are observed in the ratio of,
    Three dominant to one recessive.
The genetic constitution (genotype),
    Of the true-breeding parental generation,
    Is homozygous,
    That is a parent has either both recessive alleles,
    Or both dominant alleles.

Since the F_1 generation receives one copy of each parental allele,
    They are said to be heterozygous,
    That is they have inherited one dominant allele from one parent,
    And one recessive allele from the other parent.

The F_2 generation may inherit any combination of alleles,
    From the F_1 generation,
    And may be homozygous dominant,
    Heterozygous,
    Or homozygous recessive.

All possible gametic fusions may be represented in a Punnett square.

Most human genetic diseases are transmitted,
    As autosomal recessive traits,
    Expressed only in the homozygous state.
As no controlled crosses can be done,
    With respect to human inheritance of genetic traits,
    Analysis is by pedigree,
    That is by examining the occurrences of a particular trait,
    In family trees.

23. WITH NAMED EXAMPLES, DISCUSS HOW THE INHERITANCE OF SEX-LINKED TRAITS DIFFERS FROM THAT OF AUTOSOMAL TRAITS

(15)

Introduction

Although certain genes on the sex-chromosomes specify sexual characteristics of an individual, these chromosomes also contain genes for traits unrelated to sex. As with autosomal traits, they may be expressed in individuals who do not possess the appropriate allele to confer dominance.

RELEVANT POINTS

Autosomal recessive traits may only be expressed in the homozygous recessive condition,
> Where an individual carries both alleles for that trait,
> The expression of which shows no selectivity,
> Between male and female offspring.

The trait may be inherited on any chromosome,
> Except the sex-chromosomes.

For an offspring to express an autosomal recessive trait,
> Both parents must carry at least one recessive allele for that trait.

Examples of autosomal recessive conditions include,
> Alkaptonuria, phenylketonuria and albinism.

Sex-linked traits are so called because inheritance is linked with that of the sex of the offspring.

Female offspring inherit an X-chromosome,
> From both parents,
> And must therefore be homozygous for a recessive trait,
> For this to be expressed in the phenotype.

Male offspring inherit an X-chromosome,
> From the mother,
> And a Y-chromosome,
> From the father.

The human X-chromosome is much larger than the Y-chromosome,
    So that males will have alleles present on the X-chromosome,
    Expressed in the phenotype,
    Whether they are dominant or recessive,
    As there is no allele present,
    To confer dominance.

Sex-linked traits therefore are predominant in males,
    And may only be inherited on the X-chromosome.
For a male offspring to express an X-linked recessive trait,
    The mother must have at least one recessive allele for that trait.
For a female offspring to express an X-linked recessive trait,
    The father must express the trait.

In the case of haemophilia,
    A female haemophiliac must have a haemophiliac father,
    And a mother who carries at least one recessive allele for this condition.

Other examples of X-linked traits are colour-blindness and Duchenne-type muscular dystrophy.

24 DISCUSS HOW GENETIC ABNORMALITIES CAN RESULT IN CERTAIN HUMAN CONDITIONS
(15)

Introduction

Alterations in the DNA sequence and gross chromosomal abnormalities at the molecular and cellular level can have extremely deleterious effects on the developing embryo. DNA analysis and karyotyping allows the early detection of certain conditions.

RELEVANT POINTS

Genetic abnormalities can arise through mutation in sections of the DNA,
    Altering the base type,
    And affecting the sequence of amino acids.
Sickle-cell anaemia arose through the mutation of a single base pair.

Other abnormalities can arise through chromosomal aberrations.
Polyploidy is the presence of multiple members of haploid (n) chromosome sets,
    A phenomenon rare in animals,
    And fatal in humans.
Triploidy (3n) is sometimes found in spontaneously aborted embryos.
Commonly genetic abnormality in humans,
    Is characterised by the presence of an extra chromosome,
    Or absence of a chromosome.
Such conditions arise as a result of abnormal meiotic (or meiotic) division,
    In which homologous chromosomes fail to separate,
    A phenomenon known as non-disjunction.
Chromosomal non-disjunction may occur during either the first or second meiotic divisions,
    Or during the subsequent mitotic division of the zygote,
    Which may lead to a clone of abnormal cells,
    In an otherwise normal individual.
Down's syndrome is caused by an extra chromosome number 21,
    Due to meiotic non-disjunction of the ovum.
Persons affected have abnormalities of the face, tongue, hands and other parts of the body,
    And are mentally retarded to varying degrees.

The occurrence of Down's syndrome correlates with maternal age.

Turner's syndrome is the only known viable human condition,
    In which a chromosome is missing.
These XO individuals have only one X-chromosome,
    Due to meiotic non-disjunction of the egg or sperm,
    And are phenotypically female.
These individuals have immature sex-organs,
    Are sterile,
    And fail to develop secondary sexual characteristics,
    But are not usually mentally retarded.

Klinefelter's syndrome is also caused by meiotic non-disjunction of the egg or sperm,
    And is characterised by an extra X-chromosome in the male.
Such XXY individuals,
    Have male sex organs,
    But testes which are abnormally small.
The syndrome often leads to formation of breasts,
    And other feminine characteristics.
Affected individuals are sterile,
    And usually of normal intelligence.

25 BY MEANS OF EXAMPLES YOU HAVE STUDIED, DISCUSS GENETIC CONDITIONS OF MEDICAL IMPORTANCE WITH REFERENCE TO THE FOLLOWING:
  (a) family history; (6)
  (b) use of karyotypes; (5)
  (c) post-natal screening. (4)

Introduction

Modern methods of detecting genetic conditions and related counselling allow couples to make informed decisions about family planning. These methods are also invaluable in the study and treatment of genetic disorders.

RELEVANT POINTS

(a) Family history

The study of family histories can determine the genotypes of individuals,
  As well as providing a preventive approach to genetic disorders.
The basic principles of determining genotype make use of the following,
  The genotype of a homozygous recessive,
  Can be displayed through their phenotype,
  While heterozygous and homozygous dominants,
  Are usually phenotypically the same,
  But their genotypes may be deduced,
  By reviewing the phenotypes,
  Of other members of the same family tree.
This process can be used to determine recessively inherited conditions,
  Such as cystic fibrosis,
  Tay-Sachs disease,
  Or dominantly inherited conditions which are expressed later in life,
  Such as Huntington's chorea.

(b) Use of karyotypes

Foetal cells may be extracted from the amniotic fluid surrounding the foetus,
  Between 14 and 16 weeks of pregnancy,
  A process known as amniocentesis.
These cells are chemically stimulated to divide,
  And then arrested in early mitosis,
  Allowing chromosomes to be stained,
  And photographed.

Such photographs may be arranged in sequence,
> To produce a karyotype,
> Allowing the detection of gross chromosomal abnormalities,
> Such as Down's syndrome,
> Which is now routinely tested for,
> In pregnant women over the age of 35.

(c) Post-natal screening

Some genetic disorders result in the alteration of metabolism,
> So that abnormal metabolic products may appear,
> In the blood,
> Or urine,
> But can be tested for in the newborn.

A routine test is for the recessively inherited condition,
> Phenylketonuria,
> Which results in abnormal metabolic products,
> Namely phenylalanine and phenylpyruvic acid,
> Detectable in the blood.

26 HOW IS THE MENSTRUAL CYCLE CONTROLLED BY HORMONES?

(15)

Introduction

The interaction of hormones and related feedback mechanisms which co-ordinate the menstrual cycle, prepare the uterine lining (endometrium) for implantation of an embryo. If fertilisation does not occur, menstruation marks the beginning of a new cycle.

RELEVANT POINTS

Hormones involved in the control of the menstrual cycle include,
    Pituitary hormones – follicle stimulating hormone (FSH) and luteinising hormone (LH),
    And ovarian hormones – oestrogen and progesterone.

FSH acts on the ovary,
    To stimulate maturation of the Graafian follicle,
    Causing the release of oestrogen from ovarian tissue.

Oestrogen acts on the pituitary gland,
    To stimulate the release of LH.

LH acts on the ovary,
    To bring about ovulation,
    Causing the Graafian follicle,
    To become the corpus luteum,
    Which is then stimulated in turn to secrete progesterone.

Progesterone promotes the proliferation of the uterine wall,
    And increases the blood supply,
    By enlargement of arteries.
Progesterone also acts on the pituitary,
    By inhibiting FSH secretion,
    And also oestrogen secretion.

At high levels,
> Progesterone inhibits LH secretion,
> So that the corpus luteum then stops secreting progesterone,
> Resulting in menstruation.

Oestrogen acts on the uterus,
> To promote repair of the uterine wall,
> After menstruation.

In the event of fertilisation,
> The levels of progesterone and oestrogen are maintained,
> By a hormonal "over-ride" mechanism,
> Which prevents menstruation,
> And spontaneous abortion of the embryo.

27. DISCUSS FERTILITY INTERVENTION WITH REFERENCE TO THE FOLLOWING:
    (a) **causes of infertility;** (8)
    (b) **treatment of infertility.** (7)

Introduction

Fertility intervention is characterised by factors preventing the union of sperm and ovum to produce a zygote, or inhibiting the zygote once formed from implanting in the uterine wall.

RELEVANT POINTS

(a) Causes of infertility

Female infertility can be caused by a number of factors.

A failure to ovulate,
    Usually hormonal in origin,
    Due to disorders of the endocrine system,
    Or to emotional stress.

Blockage of the uterine tubes,
    Due to infections, fibroids, cysts or cancer,
    Or to stress-induced spasms of the oviduct.

Failure to implant,
    Due to hormonal imbalance,
    Or infection.

Male infertility is usually caused by a low sperm count,
    Which may be due to a variety of causes such as,
    Hormonal imbalances, poor health or alcohol consumption.
A relationship exists between smoking and production of irregular sperm,
    And/or a low sperm count.

Infertility may also be due to exposure to mutagens,
    Ionising radiation,
    Or chemical toxins.

(b) Treatment of infertility

Infertility treatments include fertility drugs,
    In vitro fertilisation (IVF),
    And artificial insemination.

Fertility drugs are employed to correct hormonal imbalances,
    And may contain FSH and LH,
    Or factors which stimulate the secretion of FSH and LH.

Antispasmodics are given to prevent spasms in the oviducts.

IVF involves the stimulation of multiple ovulations,
    Using fertility drugs,
    The removal of a number of ova,
    Through the abdominal wall,
    Fertilisation with sperm,
    In specialised fluid medium,
    Embryonic cell division,
    And final implantation of several embryos.

Artificial insemination involves the insertion of donor,
    Or partner's sperms,
    Into the cervix.

28. DISCUSS INTRA-UTERINE DEVELOPMENT UNDER THE FOLLOWING HEADINGS:
    (a)   exchange between maternal and foetal circulations;   (10)
    (b)   influence of placental hormones.   (5)

Introduction

The passage of materials between maternal and foetal circulations provides for respiratory gas exchange and nutrient transfer as well as waste removal for the embryo. In addition the placenta produces important hormones.

RELEVANT POINTS

(a) Exchange between maternal and foetal circulations

The placenta is the organ of exchange,
    Between the maternal and foetal circulations.
Maternal and foetal blood do not usually make physical contact.
The foetus is provided with glucose,
    Which is actively taken up,
    From the maternal circulation,
    Antibodies,
    Which are passed through by pinocytosis,
    Oxygen,
    Which is transferred by diffusion.

Harmful substances may also cross the placenta to the foetus.
Such substances include teratogens,
    Such as alcohol, nicotine and heroin,
    Which may cause birth defects,
    Mutagens,
    Such as ionising radiation,
    Which may cause genetic defects,
    Pathogens,
    Such as rubella and HIV,
    Which may cause damage,
    Or foetal infection.

Foetal metabolic wastes,
    Such as carbon dioxide,
    Are passed from the foetal to maternal circulation.

(b) Influence of placental hormones

The placenta also serves as an endocrine organ.
Around the eleventh week of pregnancy,
    The placenta secretes oestrogen and progesterone,
        Previously secreted by the ovary and corpus luteum respectively,
    The latter now degenerate.
These placental hormones continue to inhibit,
    Ovulation and menstruation,
    But stimulate development of milk producing tissues of the mammary glands.
The pituitary hormone prolactin,
    Is secreted towards the end of gestation,
    And stimulates lactation,
    In the mammary glands.

29 DISCUSS BIRTH WITH REFERENCE TO EACH OF THE FOLLOWING:
  (a) induction of birth; (7)
  (b) nutrition of the newborn. (8)

Introduction

The factors which initiate the process of birth (parturition) are not fully understood but are known to involve certain hormonal interactions. Human milk production and ejection are also under hormonal control.

RELEVANT POINTS

(a) Induction of birth

During pregnancy,
    Progesterone,
    Secreted by the placenta,
    Inhibits muscle contraction of the uterine wall.
Towards the end of pregnancy,
    Progesterone levels fall,
    While the level of oestrogen,
    Secreted by the placenta,
    Increases,
    Which sensitises the uterine muscles,
    To the effect of the pituitary hormone,
    Oxytocin.

Oxytocin stimulates the uterine muscles to contract,
    Causing "labour" to start,
    Which can also be induced,
    By the deliberate injection of artificial hormones,
    Which mimic the effect of natural oxytocin.

(b) Nutrition of the newborn

For the first few days after birth,
    The mammary glands produce a fluid,
    Called colostrum,
    Which provides a concentrated source of antibodies,
    Giving passive immunity to the newborn infant.

Usually by the third day,
    Breast milk is produced.
Colostrum and milk,
    Provide all the nutrients needed by the newborn.
Human milk contains less protein and minerals,
    But more lactose,
    Than modified cow's milk,
    Which is also more likely to produce,
    Allergies towards dairy products.

Breast-feeding,
    Promotes the recovery of the uterus,
    Reduces the development of infantile diarrhoea,
    And respiratory infection,
    During the second six months of life.

Toxic substances can be excreted in milk,
    And ingested by the infant.
Such substances include,
    Organochlorine pesticides,
    Mercury and lead compounds,
    Alcohol, nicotine and "hard" drugs.

30. DESCRIBE THE PATTERN OF GROWTH AFTER BIRTH UNDER THE FOLLOWING HEADINGS:

(a) major stages of growth; (5)
(b) role of growth hormones; (5)
(c) hormonal changes and development at puberty. (5)

Introduction

The control of growth and development involves the interaction between various endocrine structures to provide co-ordinated hormonal action and may be affected by external and environmental factors.

RELEVANT POINTS

(a) Major stages of growth

The rate of growth is shown by characteristic "growth curve",
    Which varies with age,
    For example, in the teenager stage,
    Growth "spurts" are evident.

The relative body proportions vary throughout development,
    From foetus to adult,
    The head size in particular,
    In the early stages,
    Is much larger in relation to the rest of the body.

Peak growth times in males and females differ,
    With high growth rates for females,
    In the early teens,
    But for males occurring,
    In the mid-teens.

(b) Role of growth hormones

Growth hormone (GH),
    Is secreted by the pituitary gland,
    Under the control of the hypothalamus.

GH stimulates growth,
> Through direct and indirect actions on different tissues,
> For example stimulation of skeletal,
> And soft tissue growth,
> And of lipid metabolism.

During childhood,
> Insufficient GH leads to hypopituitary dwarfism,
> While overproduction leads to gigantism.

In adulthood,
> Overproduction leads to acromegaly.

(c) Hormonal changes and development at puberty

Changes at puberty,
> Are initiated by the hypothalamus,
> With its hormonal stimulation of the pituitary gland,
> Resulting in secretions by the pituitary,
> Of gonadotrophin hormones,
> Such as luteinising hormone (LH),
> And follicle-stimulating hormone (FSH).

The gonads respond to gonadotrophins,
> By increased secretions of oestrogen,
> And also progesterone (from ovaries) in the female,
> Or testosterone (from testes) in the male,
> Causing the development of secondary sexual characteristics.

31. DESCRIBE IN DETAIL THE STRUCTURES THROUGH WHICH OXYGEN AND CARBON DIOXIDE WILL PASS AS THEY ARE TRANSPORTED BETWEEN THE ALVEOLI IN THE LUNGS AND RESPIRING CELLS IN THE BRAIN

(15)

Introduction

Gaseous exchange takes place in the lungs with carbon dioxide being carried from the brain in the venous system and expelled from the blood into the alveoli and oxygen being taken into the blood from the alveoli and carried in the arterial system to the brain.

RELEVANT POINTS

- Oxygen diffuses across the one-cell thick alveolar wall,
  - Into the blood carried by the pulmonary vein,
  - To be transported back to the left atrium of the heart,
  - Passing through the bicuspid valve,
  - Into the left ventricle,
  - Passing through the aortic semi-lunar valve,
  - Into the aorta,
  - Carotid and vertebral arteries supplying the brain,
  - Finally diffusing out from capillaries,
  - To respiring brain cells.

- Carbon dioxide diffuses from respiring brain cells,
  - Into the blood carried by the jugular and vertebral veins,
  - To be transported via the superior vena cava,
  - To the right atrium of the heart,
  - Passing through the tricuspid valve,
  - Into the right ventricle,
  - Passing through the pulmonary semi-lunar valve,
  - Into the pulmonary artery,
  - To the lungs where diffusion takes place in the alveoli,
  - Expelling carbon dioxide to the outside.

32 HOW IS BLOOD PRESSURE IN VESSELS CAUSED AND HOW DOES IT CHANGE IN DIFFERENT PARTS OF THE CIRCULATION?

(15)

Introduction

Blood is under changing pressure in the vessels in different parts of the body. This pressure is a combination of the contraction of the left ventricle pumping blood into the arterial system (systolic pressure), the recoiling effect of the muscular walls of the large arteries as the left ventricle fills (diastolic pressure) and the resistance encountered as the blood flows through vessels.

RELEVANT POINTS

Blood pressure is the force exerted by circulating blood on the walls of the containing vessels
    Caused by a combination of the force exerted by the contraction of the heart,
    And the resistance offered by the vessels as the blood passes through.

Blood flows through the transport system because of pressure differences in various parts of the body,
    The flow of blood will be from the area of high pressure in the aorta (equivalent to 100mm Hg),
    To the area of low pressure in the right atrium (equivalent to 0mm Hg).

As blood flows through arteries, arterioles and capillaries, the pressure falls,
    Proportional to the frictional resistance in the vessels,
    Which is relatively little in the large and medium sized arteries,
    Since they are near the heart and have large cross-sections,
    But relatively high in the narrow arterioles which,
    Offer more than half the total resistance to blood flow,
    With decreasing pressure in the capillaries which,
    Although they are narrower than arterioles,
    Are present in much larger numbers than arterioles,
    And therefore have a greater total cross-sectional area.

Resistance to blood flow will increase in proportion to the length of the conducting vessel,
> A long vessel offering more resistance,
> A short vessel offering less resistance.

Changes in the diameter of vessels,
> Due to vasoconstriction or vasodilation,
> Will also affect arterial blood pressure.

Vasoconstriction of arterioles,
> Causes an increase in frictional resistance,
> And a corresponding increase in arterial blood pressure.

Vasodilation of arterioles causes the opposite effect.

The location of the blood vessel in relation to the heart also affects blood pressure,
> Thus when standing upright,
> A vessel below the level of the heart has an elevated pressure,
> While a vessel above the level of the heart has a decreased pressure,
> Due to the effect of gravity.

33 *AS WELL AS CIRCULATING FLUID IN BLOOD VESSELS, THE BODY CIRCULATES FLUID IN LYMPH VESSELS. GIVE AN ACCOUNT OF THIS "SECONDARY CIRCULATORY SYSTEM" UNDER THE FOLLOWING HEADINGS:*
- (a) **formation and importance of lymph;** (4)
- (b) **lymph circulation;** (7)
- (c) **role of lymph nodes in defence.** (4)

Introduction

The lymphatic system performs various functions including the absorption of excess tissue fluid and the transport of this fluid, now called lymph, back into the general circulation. Lymph nodes are involved in helping the body's defence against possible pathogenic attack.

RELEVANT POINTS

(a) Formation and importance of lymph

Lymph is formed when excess tissue fluid,
    Is absorbed into lymphatic vessels.
Lymph restores to the blood small amounts of proteins,
    Which are forced across the capillary walls into the tissue fluid.
Digested fats absorbed from the small intestine,
    Into the lacteals,
    Are carried in lymph.

(b) Lymph circulation

Lymph travels in thin-walled, blind-ending, lymphatic vessels,
    Found in all body tissues,
    Which empty their contents,
    Via the thoracic and right lymphatic ducts,
    Into the venous blood supply,
    Thus returning excess fluid,
    From extracellular spaces,
    Back to the general circulation,
    To maintain blood volume.

Lymph travels in one direction only,
> From body organs back to the heart,
> Using a mechanism similar to flow of blood in the venous system,
> Which does not rely on pressure generated by the heart muscle,
> But on forces generated by skeletal muscles outside the lymphatic vessels,
> And respiratory movements,
> In conjunction with one-way semi-lunar type valves.

(c) Role of lymph nodes in defence

Lymph nodes are small structures located along the length of lymphatic vessels,
> Filtering out possible pathogens,
> Such as bacteria,
> And other foreign debris,
> Which are then phagocytosed by macrophages.

Lymph nodes also produce lymphocytes,
> Which may be discharged into the lymph,
> And eventually into the blood,
> While plasma cells produce antibody molecules.

34
RED BLOOD CELLS ARE AMONGST THE MOST UNUSUAL AND PLENTIFUL CELLS IN THE HUMAN BODY. WRITE AN ACCOUNT OF THESE CELLS WITH REFERENCE TO THE FOLLOWING:
- (a) relationship between structure and function; (6)
- (b) production and eventual breakdown. (9)

Introduction

Red cells are the most common cell found in the blood, approximately 5 million cells per cubic millimetre, whose structure is very much linked to their main function of oxygen transport. They are produced in the marrow of particular areas of the skeleton and eventually destroyed when they become old and worn out.

RELEVANT POINTS

(a) Relationship between structure and function

Red cells have no nucleus,
    Or other sub-cellular organelles,
    Giving cells unrestricted space,
    To carry the respiratory pigment haemoglobin,
    Which makes up about one third of the cell mass,
    For oxygen uptake.
Their small size (about 8μm in diameter, 2·2μm thick at their edge and 1μm at their centre),
    And biconcave shape,
    Present a very large surface area in relation to volume,
    For oxygen uptake.
The membrane is flexible,
    Allowing the cells to squeeze through narrow capillaries unhindered.

(b) Production and eventual breakdown

Red cells are produced in red bone marrow of skull, vertebrae, ribs, pelvis,
    And the ends of the long bones,
    Such as the femur and humerus.

Red cells originate from unspecialised stem cells in the bone marrow,
    Potentially at a very high rate of about 2·5 million cells per second.

In order to maintain the red cell count,
    The rate of formation must balance the rate of loss,
    This rate, which is essentially subject to feedback control,
    Being indirectly stimulated by the oxygen levels in the blood,
    Which, when low, cause an increase in the rate of cell production,
    And when high, cause a decrease in the rate of cell production,
    Thereby maintaining blood viscosity and homeostasis.

Red cells, being enucleate, have a limited life span,
    Of the order of 120 days,
    Worn out red cells may be phagocytosed in the spleen, liver and bone marrow,
    Or their fragile membranes simply burst when exposed to physical forces as the cells circulate.
The haemoglobin is broken down to haem and globin,
    With the globin and iron from the haem being recycled in the marrow,
    And the remainder being excreted by the liver,
    As a bile pigment called bilirubin.

Normal development of red cells also is influenced by dietary intake,
    Requiring adequate amounts of vitamin B_{12} and iron.

Red blood cell production is stimulated by a hormone,
    Called erythropoietin,
    Released by the kidneys.

35 *DISCUSS THE ROLE OF THE LIVER UNDER THE FOLLOWING HEADINGS:*
   (a) **metabolism of protein;** (5)
   (b) **detoxification;** (5)
   (c) **conservation of useful substances.** (5)

Introduction

The liver is the largest gland in the body weighing over a kilogram with a range of complex functions and an important role in body homeostasis. Its functions include the metabolism of protein, detoxification of harmful substances such as drugs and poisons and the conservation of useful materials such as carbohydrate.

RELEVANT POINTS

(a)   Metabolism of protein

Amino acids from protein digestion
   Are absorbed into the bloodstream and pass through the liver,
   Possibly to be used in the synthesis of proteins,
   Such as blood clotting proteins,
   While excess amino acids are deaminated in the liver,
   By the removal of the amino group $-NH_2$,
   To be converted to ammonia,
   And non-essential amino acids,
   Which can be transaminated in the liver,
   Into other amino acids.

Plasma proteins,
   Such as clotting factors and albumin,
   Are also synthesised in the liver,
   And once they are no longer able to perform their function,
   Are deaminated in the liver.

A major bile pigment bilirubin,
   Is a breakdown product of the metabolism by the liver,
   Of the protein haemoglobin.

(b) Detoxification

When proteins are metabolised in the liver,
    Ammonia is formed,
    Which is highly toxic,
    But rendered harmless by liver cells,
    Which convert the ammonia into urea,
    With a low toxicity,
    To be safely excreted in the urine.

Hormones which have fulfilled their functions,
    Are broken down in the liver into harmless substances,
    Which may be excreted in bile or in the urine.

Drugs such as antibiotics,
    Can be inactivated by chemical modification in the liver.

The liver also has fixed macrophages,
    Which ingest and digest,
    Antigens.

(c) Conservation of useful substances

One of the most important functions of the liver is to conserve excess glucose,
    By storing it as glycogen,
    Or converting it to fat,
    Until it is required.

Iron from worn out red blood cells,
    Is conserved in the liver to be used in the synthesis of more haemoglobin,
    By first becoming attached to a plasma protein called transferrin,
    And then being transported in the blood to bone marrow where synthesis occurs.

Iron from worn out red cells,
    May also be conserved in the liver,
    In the form of chemical substances,
    Such as ferritin.

The liver also stores vitamins A, B_{12}, D, E and K,
    And minerals such as iron and copper.

36 GIVE AN ACCOUNT OF THE ROLE OF THE KIDNEYS WITH REFERENCE TO THE FOLLOWING:
   (a)   ultrafiltration;   (8)
   (b)   reabsorption.   (7)

Introduction

The kidneys are important organs of homeostasis. By removing wastes and conserving useful materials, they prevent the build-up of potentially harmful substances and preserve the balance of the osmotic potential of the blood.

RELEVANT POINTS

(a)   Ultrafiltration

Blood pressure in the large arteries is of the order of 100mm Hg,
   And falls to about 50mm Hg in the capillaries of the glomerulus,
   Which is very much higher than other body capillaries,
   So that fluid is forced out into the Bowman's capsule,
   And simultaneously subjected to ultrafiltration in the glomerulus,
   Since the capillaries of the glomerulus are freely permeable to water and small molecules,
   Such as salts, glucose, amino acids, vitamins and urea,
   But impermeable to large molecules,
   Such as plasma proteins,
   Giving rise to glomerular filtrate.

Unlike reabsorption,
   Ultrafiltration is a relatively non-selective process,
   So that the glomerular filtrate is almost identical to blood plasma,
   Apart from not containing plasma proteins.

Each kidney contains over a million nephrons,
   Allowing approximately 0·5ml of urine per minute to be produced by each kidney,
   And it is estimated that the total blood volume of about 5·5 litres,
   Is ultrafiltered by the kidney nephrons once every forty minutes.

(b) Reabsorption

More than half of the selective reabsorption from the kidney tubule,
>Takes place into the capillaries which are in close contact with the proximal part of the tubule,
>Thus preventing the loss of useful substances,
>Such as water, glucose, amino acids and vitamins,
>Which are secreted back into the bloodstream,
>Against a concentration gradient,
>Requiring energy in the form of adenosine triphosphate.

The kidneys are able to alter the rates of reabsorption and excretion,
>In response to differing situations,
>Thus helping preserve homeostasis,
>By being subjected to physiological controls,
>Such as the hormone anti-diuretic hormone.

37 WITH REFERENCE TO THE MAINTENANCE OF BLOOD SUGAR LEVEL, EXPLAIN WHAT IS MEANT BY NEGATIVE FEEDBACK CONTROL

(15)

Introduction

Many of the body's physiological homeostatic mechanisms are subjected to feedback control which allows responses to be altered to meet changing needs. When the control results in an increasing stimulus causes a decreasing physiological response, or vice-versa, it is called negative. For example, when the blood sugar level falls below or rises above its normal threshold, mechanisms operate to increase or decrease the blood level respectively.

RELEVANT POINTS

Glucose is the body's main energy source,
    And it is vitally important its concentration in the blood be maintained within narrow limits.

The main hormones regulating blood sugar levels are insulin, glucagon and adrenaline.
    Insulin and glucagon,
    Are produced respectively by the α and β cells of the islets of Langerhans in the pancreas,
    While adrenaline,
    Is produced by the adrenal glands.

The secretion of insulin and glucagon is almost entirely dependent upon the concentration of glucose in the blood,
    With the α and β cells in the islets of Langerhans,
    Able to detect and react to the changing glucose concentrations,
    Thus regulating blood sugar level via negative feedback control.

When blood sugar level is high,
    The β cells of the islets of Langerhans are stimulated,
    To release insulin into the blood,
    To affect special receptors in target cells,
    Causing them to uptake glucose more readily,
    While inhibiting the release of glucose from the liver,

And stimulating the liver to convert glucose to glycogen,
Which is insoluble and osmotically inert,
Thus restoring the blood sugar level back to normal.

When blood sugar level is low,
The α cells of the islets of Langerhans are stimulated
To release glucagon into the blood,
Which stimulates the liver to convert glycogen into glucose,
And inhibits the synthesis of glycogen,
Thus restoring blood sugar level back to normal.

Adrenaline also participates in regulating blood sugar level,
In a similar way to the effects of glucagon,
By stimulating the breakdown of glycogen in the liver,
To form glucose,
Which is then released into the blood,
Raising the blood sugar level.

38 DISCUSS PHYSIOLOGICAL REGULATION WITH REFERENCE TO THE FOLLOWING:
    (a)   control of heart rate;   (6)
    (b)   effects of exercise on the cardiovascular system.   (9)

Introduction

The heartbeat varies to meet changing demands, beating faster for example during exercise to supply oxygenated blood to skeletal muscles and more slowly during sleep when the demand for oxygenated blood is less.

RELEVANT POINTS

(a) Control of heart rate

Only one region of the heart is capable of originating the nerve impulses needed to stimulate the muscle of the wall of the heart,
    This is the pacemaker or sinoatrial node,
    Situated in the right atrium of the heart,
    Near the point where the superior vena cava enters.

During systole,
    The sinoatrial node generates nerve impulses,
    First stimulating nearby myocardial cells of the left and right atria to contract,
    And then passing into the myocardial cells of the left and right ventricles,
    Via specialised conducting tissue,
    Made up of the atrioventricular node,
    Located below the sinoatrial node,
    And conducting tissue,
    Located in the septum and ventricular walls,
    Causing the ventricles to contract,
    To eject their contents into the pulmonary arteries and aorta.

External control of the heart rate is via fibres of the autonomic nervous system,
    Which emerge from a special centre located in the medulla,
    Sympathetic stimulation causing an increased heart rate,
    Parasympathetic stimulation causing a decreased heart rate.

Apart from external nervous control,
> The heart rate is also influenced by hormones,
> Such as adrenaline,
> Which increases heart rate,
> During emergency situations,
> Provoking what are sometimes referred to as "fight or flight" responses,
> And thyroxine which also speeds up heart rate,
> But takes several hours to do so.

(b) Effects of exercise on the cardiovascular system

During exercise the demand by skeletal muscles for oxygen can increase 100 times their normal level,
> Thus exercise causes an increased demand for oxygenated blood to skeletal muscles,
> So the heart rate and volume output must therefore rise to meet this demand,
> And can reach a rate of 195 beats per minute,
> With an output in excess of 35 litres per minute during strenuous exercise.

There are a number of changes to blood vessels during exercise,
> Arterioles supplying the exercising muscles (including the heart itself) dilate,
> Causing a decrease in frictional resistance to blood flow,
> While visceral vessels constrict causing an increased resistance to blood flow,
> Which results in a decreased blood flow to the organs of the gut,
> And increasing the available volume of blood to flow to exercising muscles.

There is also a parallel increase in the return of venous blood to the heart,
> By increased activity of "muscle pumps" acting on the valves,
> And increased respiratory movements.

39 THE TEMPERATURE OF THE BODY IS NORMALLY VERY STABLE DESPITE FLUCTUATIONS IN THE ENVIRONMENT. HOW IS THIS STABILITY ACHIEVED?

(15)

Introduction

Humans are homeothermic, maintaining a constant internal body temperature regardless of changes in the environment. This regulation is effected not only by voluntary actions but also by involuntary mechanisms which are co-ordinated by the hypothalamus.

RELEVANT POINTS

In order to maintain a constant temperature in the structures below the skin,
  That is the core temperature,
  The heat produced must balance the heat lost,
  Otherwise damage to tissues will result,
  From overheating which is fatal because of protein denaturation,
  While too low a core temperature is fatal by causing the heart beat to become irregular.
The mechanisms which attempt to maintain core temperature at a normal value of 37°C,
  Ensure that the heat production and heat loss balance,
  So that the core temperature hardly fluctuates more than 1°C from the normal value.

The main temperature regulatory centre of involuntary control is in the hypothalamus,
  Which receives information from temperature receptors located in the skin,
  And the hypothalamus itself,
  Which detects a change in the blood temperature,
  By responding rapidly when the temperature rises above normal,
  To stimulate responses which lower body temperature,
  And very slowly when the temperature falls below normal,
  To stimulate responses which elevate body temperature,
  Effectively acting as a body thermostat.

When the blood temperature falls below normal, the hypothalamus sets in motion responses which increase and retain body heat by,
> Causing peripheral blood vessels in the skin to constrict,
> Which decreases radiative, convective and conductive heat loss from the skin,
> Raising the rate of metabolism,
> Which generates more heat,
> Stimulating skeletal muscles to repeatedly contract and relax very quickly,
> Which is shivering,
> And can cause up to a four fold rise in body heat production in as many minutes,
> Increasing thyroxine production,
> Which in turn increases metabolic rate,
> Causing the surface hairs to stand erect,
> But this has little importance in humans.

When the blood temperature rises above normal, the hypothalamus sets in motion responses which promote the release of body heat by,
> Causing peripheral blood vessels in the skin to dilate,
> Which increases radiative heat loss from blood supplying deep organs,
> Stimulating sweat glands to increase their function of producing perspiration,
> Which cools the skin as it evaporates,
> Inhibiting shivering,
> Decreasing the rate of metabolism.

These involuntary actions are examples of negative feedback which help preserve homeostasis.

In addition to these involuntary actions, voluntary activities, often behavioural, are brought into play.

For example when cold,
> Curling up to expose reduced surface area,
> Increasing clothing to insulate the body,
> Taking in hot foods,
> Which decrease heat loss.

When warm a general lack of activity,
> The shedding of clothing,
> And the taking in of cold drinks,
> Which decrease heat production.

40 GIVE AN ACCOUNT OF FUNCTIONAL AREAS IN THE CEREBRUM
(15)

Introduction

The brain consists of an estimated one hundred billion neurons and some eighty per cent of these are located in the cerebrum grouped into discreet areas which perform particular functions. These areas may receive incoming sensory information and react by generating motor impulses or performing higher activities such as learning.

RELEVANT POINTS

The cerebrum is not concerned with a particular sense or function,
    But instead contains a large number of functional or association areas,
    Which receive incoming information from within and without the body,
    Process and then relay this information to other areas,
    Which respond to it.

Evidence suggests that the cortex of the cerebrum exhibits distinct localisation of function in particular well mapped out areas such as the,
    Motor areas,
    Which initiate voluntary movements in skeletal muscles,
    Sensory areas,
    Which receive information from receptors situated in the skin and elsewhere,
    Association areas,
    Which connect motor and sensory areas,
    Involved in vision, hearing, memory and judgement.

A feature of these areas is that their sizes are not uniform,
    But instead are related to their function,
    So that regions of the body which have a high density of receptor cells,
    Such as the fingertips and lips,
    Have a correspondingly large sensory area on the cortex,
    And regions of the body which have a high density of motor connections,
    Such as the hands,
    Have a correspondingly large motor area on the cortex.

Another feature of these functional areas is that some are connected to the opposite side of the body,
    So that the left cerebral hemisphere,
    Controls voluntary movements of the right side of the body and vice versa.

41 CONTRAST THE TWO DIVISIONS OF THE AUTONOMIC NERVOUS SYSTEM

(15)

Introduction

Nerve fibres which stimulate contraction of the muscles of the heart and gut and particular glands to secrete make up the autonomic nervous system. These fibres are of two types based on structural and function differences giving rise to two divisions of the autonomic nervous system, the sympathetic and parasympathetic. The autonomic nervous system plays a critical role in body homeostasis.

RELEVANT POINTS

The autonomic nervous system is a branch of the central nervous system,
    But contains only motor fibres,
    Not under the voluntary control of the cerebrum,
    Which are of two types, each belonging to one of the two divisions,
    Sympathetic and parasympathetic,
    Giving almost every body organ a double nerve supply,
    Each of which is usually functionally antagonistic to the other.

If the sympathetic branch stimulates an effector such as cardiac muscle,
    To increase rate of contraction,
    The parasympathetic branch will inhibit that muscle,
    Decreasing its rate of contraction,
    And the balancing of the action of these two,
    Allows for homeostatic regulation of heart beat,
    And other body functions.

In a similar way the autonomic nervous system has antagonistic actions on other effectors as shown in the table on the following page:

ORGANS AFFECTED	SYMPATHETIC ACTION	PARASYMPATHETIC ACTION
arteries	constricted; increased blood pressure	dilated; decreased blood pressure
gut	inhibits peristalsis	stimulates peristalsis
digestive glands	decreased secretion	increased secretion
muscles of hairs	erected	relaxed
pancreas	decreased secretion	increased secretion
bladder	relaxed	constricted
muscles of the bronchi	dilated	constricted
iris	dilated	constricted
sweat glands	increased secretion	decreased secretion

Normally the sympathetic branch brings about physiological changes which prepare the body to deal with stressful situations,
    By increasing heart beat and breathing rates,
    Whereas the parasympathetic branch operates in reverse,
    Restoring physiology back to its homeostatic norm.

The two divisions of the autonomic system also differ in their chemical transmitters,
    Acetyl choline being released by parasympathetic fibres,
    Noradrenaline being released by sympathetic fibres.

42. DISCUSS THE ROLE OF THE SYNAPSE IN THE TRANSMISSION OF NERVE IMPULSES

(15)

Introduction

Nerve fibres do not actually touch one another but instead are separated by distinct microscopic gaps called synapses. These gaps allow for control over the direction of nerve impulses and whether an impulse is allowed to pass or is inhibited. In this way, synapses exert a great deal of control over the transmission of nervous messages.

RELEVANT POINTS

In order for the nervous system to function as an integrated unit,
- There has to be connections called synapses,
- Between the nerve cells which make up the system.

Synapses are distinct spaces between two or more nerve cells,
- Of the order of 20nm,
- Such that an impulse travelling in one nerve cell,
- Influences the next nerve cell(s).

Synapses operate using chemical transmitters,
- Such as adrenaline or acetylcholine,
- Which are secreted by the end of the axon,
- Down which a nerve impulse is travelling,
- And released into the synapse where they diffuse across,
- To stimulate the next nerve cell(s) to continue the transmission of the impulse.

Once a transmitter has functioned it may be,
- Destroyed enzymatically,
- As is acetylcholine,
- Or removed by being taken back into the axon,
- To be recycled,
- As is noradrenaline.

The chemical transmission at the synapse is unidirectional,
> Because of the anatomy of the terminal end of the axon,
> Where the transmitter is stored in vesicles,
> And the anatomy of the dendrite of the next nerve cell in line,
> Where membrane receptors are located,
> Thus preventing impulses travelling along wrong pathways.

The synapse affords a mechanism for controlling the passage of a nerve impulse,
> Which is allowed to cross an excitatory synapse,
> But inhibited by an inhibitory synapse,
> Which effectively blocks transmission.

43 OUTLINE THE ORGANISATION OF THE NERVOUS SYSTEM UNDER THE FOLLOWING HEADINGS:
   (a)   central nervous system; (5)
   (b)   consequences of myelination; (5)
   (c)   plasticity of the response of the nervous system. (5)

Introduction

The central nervous system, in conjunction with other parts of the nervous and endocrine systems, co-ordinates and integrates body function. It is made up of an enormous number of nerve cells, many of which are covered in a sheath made of myelin allowing rapid transmission of impulses. Responses of the nervous system can be altered to meet changing circumstances.

RELEVANT POINTS

(a) Central nervous system

As the name suggests, the central nervous system exerts control over the whole nervous system.
All information from within and without the body must pass through the central nervous system to be,
    Interpreted,
    And stimulate appropriate responses.

The central nervous system consists of the brain and spinal cord.
The brain is considered to be divided into three functional areas,
    Cerebrum,
    Associated with higher mental activities,
    Cerebellum,
    Associated with muscular co-ordination and balance,
    Medulla oblongata,
    Associated with autonomic activity.

The spinal cord is continuous with the medulla,
    Passing down and protected by the vertebral column,
    Providing a major trunk connection,
    Between the peripheral nervous system and the brain,
    By relaying sensory information to the brain,

In ascending tracts which conduct impulses up the spinal cord,
And relaying motor impulses from the brain,
In descending tracts which conduct impulses down the spinal cord.

(b) Consequences of myelination

Some axons in the peripheral and central nervous system are covered with a sheath,
    Made of myelin and therefore called myelinated,
    Which insulates axons from each other.

There are gaps in the myelin sheath,
    Called nodes of Ranvier,
    Which are the only part of the axon left exposed,
    To produce nerve impulses,
    And which allow the transmission of nerve impulses much more rapidly than non-myelinated axons.

Myelin sheaths are produced late in foetal development,
    And during the first year of life,
    Increasing from birth to maturity.
Since the process of myelination is not complete during early life,
    An infant is not able to respond as quickly or in such a co-ordinated way as an older child.

(c) Plasticity of the response of the nervous system

The central nervous system consists of a large number of pathways,
    Which may converge,
    So that impulses may travel along several nerve fibres,
    But be modified,
    To stimulate fewer outgoing nerve fibres,
    Or diverge,
    So that impulses may travel along few nerve fibres,
    But be modified,
    To stimulate more outgoing nerve fibres,
    Giving rise to the concept of the nervous system being "plastic" and not necessarily fixed in its response.

A good example of this plasticity is the modification of reflex actions,
> Which are functional units of the central nervous system,
> Having a profound role in integrating body function,
> And contributing to homeostasis,
> Since they operate on muscle contraction, heart rate, respiration and digestion.

This modification may be brought about by the action of the cerebrum,
> Which has sensory synaptic connections with relay nerve cells in the spinal cord,
> Thereby having conscious awareness of the stimulus evoking the reflex.

Thus a person might unknowingly lift an object which is very hot,
> And override the normal reflex action to drop it,
> By sending signals from the cerebrum down the spinal cord,
> Which inhibit the completion of the reflex response.

44. LEARNING DEPENDS HEAVILY ON INFORMATION BEING ENCODED, STORED AND RETRIEVED. HOW ARE EACH OF THESE PROCESSES THOUGHT TO OCCUR?

(15)

Introduction

In order for learning to take place sensory information, such as sounds and sights, must be converted into a form which can be held in memory to be recalled later. These events are termed encoding, storage and retrieval respectively.

RELEVANT POINTS

Sensory information enters the central nervous system,
    As a series of nerve impulses,
    Which must then be converted into an appropriate form,
    For storage in memory,
    After the original stimuli have ceased,
    A process called encoding.
Encoding involves a selective focusing on sensory information,
    To the exclusion of other stimuli,
    A process called attending,
    Lack of which leads to difficulties in remembering.

Encoding can take various forms,
    Such as a representation of the sound,
    Called acoustic encoding,
    Or a representation of the sight,
    Called visual encoding,
    Or a representation of the meaning,
    Called semantic encoding.
Evidence suggests that acoustic encoding is particularly dominant in short-term memory,
    But that semantic encoding is dominant in long-term memory,
    Especially in encoding verbal items.

Storage of information in short-term memory,
  Is extremely limited,
  Of the order of five to nine items,
  A phenomenon known as memory span,
  But can be affected by information stored in long-term memory,
  To reorganise items in short-term memory,
  Into distinct units, called "chunks",
  Which allows the short-term memory potential to be increased.
Information in short-term memory has only a limited life,
  And decays rapidly,
  As it gives way to new information coming in.
However information in short-term memory may be transferred into long-term memory,
  By rehearsing, organising or elaborating the meaning of the item(s).
Evidence suggests that the ability to retrieve information from short-term memory,
  Depends on a sequential examination of the items,
  And therefore slows down as the number of items stored increases.
Retrieval from long-term is not dependent on the number of items,
  But on an efficient accessing system to the stored information,
  So that if the original information is located in "the wrong place",
  It will not be accessible and therefore will be "forgotten",
  As shown in experiments with subjects under hypnosis,
  Who can remember experiences in childhood which they could not when conscious.
Retrieval also depends to some extent on cues,
  Which are signals eliciting long-memory,
  The better the retrieval cue,
  The better the long-term memory response.

45 OUTLINE THE EVIDENCE FOR THE MOLECULAR BASIS OF MEMORY

(15)

Introduction

Memory fundamentally requires the learning of new information, its storage and subsequent retrieval. What connects these processes in a physical way, is an alteration or alterations in the properties of the brain which may be associated with biochemical changes. This is sometimes referred to as the molecular basis of memory.

RELEVANT POINTS

Research into the molecular aspects of memory,
    Has used a number of strategies in conjunction,
    Such as comparing the brains of animals which have learned,
    With animals which have not,
    Or by looking at chemicals in humans and animals,
    Which improve or inhibit learning.
It is extremely demanding to construct properly controlled experiments into the molecular basis of memory,
    Because of the difficulty of separating biochemical events which may occur in the brain during learning,
    From other phenomena which may take place simultaneously.

It is known that certain chemicals which interfere with normal brain function,
    Will affect short-term memory function,
    Perhaps even destroying all the items in temporary storage there.
Other chemicals may actually improve short-term memory function,
    As well as the transfer of information from short to long-term memory.
Chemicals which speed up protein synthesis,
    Are known to enhance learning,
    While those which inhibit protein synthesis,
    Have little effect on short-term memory function,
    But prevent the proper development of long-term memory.

Experiments involving the injection of minute quantities of such chemicals into the brain,
> Suggest that the limbic system is particularly important in both,
> The function of short-term memory,
> And the transfer process to long-term memory.

Since long-term memories are fixed,
> It is suggested that they are in fact encoded by means of physical changes in brain structure,
> In terms of altered nerve pathways and synaptic connections,
> Including, for example, increases in the number of storage vesicles containing transmitter substances,
> Such as acetylcholine and glutamate.

The development of Alzheimer's disease,
> In which memory function is seriously affected,
> Is known to be associated with the loss of those brain cells which produce acetylcholine,
> A reduction in which seems linked to memory failure.

The neurotransmitter glutamate is particularly present in the nerve cells in the limbic system,
> Where NMDA receptors,
> From the chemical N-methyl D-aspartate which is used to detect it,
> Are found in large numbers,
> Supporting the view that the NMDA receptors are important in memory formation.

Biochemical studies have so far failed to demonstrate the presence of actual "memory molecules",
> But instead show more general changes in the brain,
> So that specific memories rely on which nerve cells are involved,
> Their location in the brain,
> And their interconnections with other nerve cells,
> Rather than particular biochemical events or substances.

46 BY MEANS OF A SUITABLE EXAMPLE, SHOW HOW MATURATION CAN INFLUENCE THE DEVELOPMENT OF BEHAVIOUR

(15)

Introduction

The process of reaching full potential in terms of growth, function or behaviour is called maturation. It is almost impossible to detach the effects of the genotype and environment in this process but the stages in the development of motor skills associated with walking are heavily influenced by maturation.

RELEVANT POINTS

Since all infants are not subjected to the same style of upbringing,
    Yet all pass through the same basic stages of motor development,
    There must be a pre-determined sequence of events,
    Which controls this development,
    And is relatively independent of environmental influences.
This process is called maturation,
    Which is important in understanding the effect of the genotype,
    On human behaviour.
The behavioural phenotypic expression of the genotype may take place,
    In stages corresponding to the time when the body is mature enough.

As an infant grows, body structures associated with walking,
    Such as bones, muscles and nerves,
    Develop and mature,
    To allow increasing co-ordination,
    In an ordered chronological sequence of events.

At birth and in the first few months,
    The infant can roll over.
By the fourth month,
    Some weight-bearing on the skeleton of the legs is possible.

By six to nine months,
> The infant can self-support,
> Sitting up alone,
> Or standing holding onto a fixed object.

By about twelve months,
> The infant can crawl about.

Finally, about fourteen months,
> The infant can stand alone,
> And walk unaided.

Until myelination is complete,
> Somewhere within the first year of life,
> Muscle response is slow,
> And co-ordination is still poor.

47 DISCUSS THE ROLE OF INHERITANCE IN BEHAVIOURAL DEVELOPMENT. YOUR ANSWER SHOULD INCLUDE INFORMATION ABOUT:
(a) role of the genotype; (8)
(b) a selected inherited disorder. (7)

Introduction

As well as controlling physical phenotype, the genotype also affects the development of behavioural phenotype. A genotype which is defective in some way, may produce an inherited behavioural disorder such as Huntington's chorea.

RELEVANT POINTS

(a) Role of the genotype

Like many human traits, behaviour is probably polygenic,
    That is, affected by many genes,
    But heavily influenced by the environment.

Much of the data on the role of inheritance has come from selective breeding experiments,
    In which animals with a particular behavioural trait,
    Such as maze-running ability,
    Are bred in various combinations.
The data obtained suggests some behaviours in animals,
    Such as aggressiveness in fowl,
    Phototaxic responses in *Drosophila*,
    Are inherited,
    Supporting the role of the genotype in determining behaviour.

Another source of data is the use of twin-studies,
    To get round the ethical impossibility of conducting human breeding experiments.
Since monozygotic twins are genetically identical,
    They also demonstrate similar inheritable behaviour patterns,
    And are similarly at risk to behavioural disorders
    Such as Huntington's chorea.

Twin studies have revealed that although the twins both inherit the same genetic disposition to disorders such as schizophrenia,
> The actual expression of the genes may take place in one twin,
> But not the other,
> Supporting the view that gene action,
> Is also influenced by environmental factors.

(b)   A selected inherited disorder

Although most human inherited traits are polygenic,
> There are some significant exceptions,
> Such as Huntington's chorea,
> Causing a deterioration in the nervous system,
> With associated behavioural dysfunction,
> And memory loss.

In the early stages of development of the disease,
> Sufferers exhibit uncontrollable motor movements,
> With these becoming more overt as the disease progresses,
> Affecting particularly the face, tongue and arms,
> Although all body parts eventually become involved,
> Leading to eventual death.

Huntington's chorea is caused by a defective autosomal dominant gene,
> Which has still not been isolated,
> But has been shown to be found on a particular chromosome,
> So that people who might be at risk,
> Can now undergo genetic testing to show whether they are likely to develop the disease,
> Around the age of thirty to forty years.

48 *HOW DOES THE DEVELOPMENT OF INTELLIGENCE DEMONSTRATE THE INFLUENCE OF A COMBINATION OF FACTORS? YOUR ANSWER SHOULD INCLUDE INFORMATION ABOUT:*
- (a) maturation; (3)
- (b) polygenic inheritance; (6)
- (c) environment. (6)

Introduction

The notion of intelligence has changed over the years, being intimately linked with the strategies for its measurement. It is generally agreed however that intelligence is the ability to learn from experience and adapt to the environment. This ability changes as a person develops and is both genetically and environmentally influenced.

RELEVANT POINTS

(a) Maturation

Intelligence reaches its "adult" level,
    When there is decrease in its measured development,
    Or when the annual increase is significantly small,
    When compared statistically with the population as a whole.
Measurable intelligence increases from infanthood,
    To reach a maximum,
    Around adolescence,
    About thirteen to fifteen years of age.

(b) Polygenic inheritance

Humans differ intellectually,
    And it is generally agreed that part of the cause of this difference,
    Is genetically based,
    And due to a combination of a large number of genes,
    That is, polygenic.

The correlation between genotypes,
    And intelligence is sufficiently high,
    Particularly in monozygotic twins,
    To support the role of the genotype.
It has been estimated variously that the proportion of intelligence due to inheritance,
    Using a score called the heritability ratio,
    Is of the order of ten to ninety percent,
    Indicating the difficulty associated with attempting to control such investigations.
Advances in the field of molecular genetics may change these views,
    If particular genes can be shown to be markers for intelligence.

(c) Environment

Although there is a high correlation between genotype and intellectual phenotype,
    Environment is also highly influential,
    As evidenced by monozygotic twins,
    Whose genotype is identical,
    Being reared apart,
    And showing different measured intelligence,
    Or genotypically different children being reared in the same environment,
    Having similar measured intelligence.
Factors such as nutrition, home environment, early intellectual stimulation,
    Will affect the full expression of the genetic potential for intelligence.
For example, studies have shown that early intellectual stimulation,
    Can have a significant impact on later performance in school.

49 WHY IS THE PERIOD OF INFANT DEPENDENCE SO LONG IN HUMANS?

(15)

Introduction

The first few years of life are usually thought of as infancy. During this time the infant, who is relatively helpless and highly dependent on parental support for a long period of time, has many opportunities for learning.

RELEVANT POINTS

From a very early age,
    An infant will display behaviour patterns,
    Such as smiling,
    Even if born blind,
    And show preferences for particular persons,
    Usually the parents,
    By audible and visual displays,
    When they see or hear these persons,
    Which reinforce parental bonding.

Memory is also developing from an early age,
    Allowing infants to compare present and past events,
    To learn, for example, that unusual objects and strangers in the environment,
    Are probably not harmful,
    While parental absence,
    Is usually only temporary,
    And need not cause anxiety.

Other mental developments are also taking place during this long period of dependency such as,
    An increasing field of view,
    The ability to distinguish colours,
    And more accurate judgement of distances.

Self-confidence also develops during the period of dependency,
    So that the infant can make decisions,
    By for example operating a particular toy,
    Or lifting something to eat,
    Without parental intervention.

Language becomes increasingly used,
    To communicate needs and emotions,
    First by incoherent sounds,
    Then by properly constructed words and language,
    So that a two year old child,
    May have a vocabulary in excess of five hundred words.

An infant will also tend to become close to particular persons,
    A phenomenon called attachment,
    Which helps prevent the young child from straying too far from its "care base".
Experiments with young children have shown,
    In the presence of such attached persons,
    Exploration of the environment,
    Play activities,
    And other mental stimulations,
    Which encourage learning
    Are much more liable to occur.

In a similar way, social interactive behaviours,
    Such as developing relationships,
    Co-operating with others,
    Are more readily learned,
    In the presence of such attached persons.

50 WHY IS NON-VERBAL COMMUNICATION SO IMPORTANT IN ADULT COMMUNICATION?

(15)

Introduction

Any communication which is not linguistically based, may be termed non-verbal. These include so-called "body language" or paralinguistics and associated movements which can convey a great deal of information without the use of spoken language.

RELEVANT POINTS

Non-verbal communication is a broad term which includes,
    Facial expressions,
    Hand and arm movements,
    Body posture,
    All potentially capable of communicating information to a receiver.
The absence of non-verbal communication,
    Is particularly evident when the sender and receiver,
    Are not in visual contact,
    As for example the difference in effectiveness,
    Of communicating by telephone,
    Or in person.

The expression on the face is perhaps the clearest mode of communicating emotion so that looking,
    Frightened,
    By raising the eyebrows,
    Furrowing of the forehead,
    May alert others to danger,
    Angry,
    By lowering and drawing together the eyebrows,
    Clenching the teeth,
    May warn of the onset of aggressive behaviour.
Facial expressions are found universally,
    And are in general not a function of a person's place of origin,
    Suggesting perhaps they might be genetically based,
    And the product of evolutionary selection.

Associated with the spoken word,
    But not language based,
    Is the tone of voice used,
    And noises,
    Including laughter, yelling or screaming.
Such "nonlexical" expressions, like facial expressions,
    Are found widely in humans from different parts of the world,
    Perhaps being nearer animal sounds,
    Which may also have a genetic component.

The timing of the delivery of the spoken word,
    Can also be a strong non-verbal cue,
    As for example reducing the number of pauses,
    To prevent another individual interrupting.

So-called eye-contact,
    Is a powerful non-verbal cue,
    The type and length of which,
    Are often used in making judgements about others.
Individuals who can sustain eye-contact,
    Usually feel close to each other.
Eye-contact can also express emotion,
    For example when the pupil is dilated or constricted.

Particular gestures may communicate information to others,
    Such as nodding of the head,
    To indicate general agreement,
    Or overt hand movements,
    To support the spoken word.

51. LANGUAGE ALLOWS THE TRANSFER OF INFORMATION THUS ACCELERATING LEARNING AND INTELLECTUAL DEVELOPMENT. DISCUSS THIS STATEMENT

(15)

Introduction

The use of language from a very early age, is often taken for granted with little or no thought to its functions. In fact, the constant contact with language can make it difficult to detach oneself enough to appreciate the complexity of this symbol system and the extent to which language serves to enhance learning and development.

RELEVANT POINTS

Language, both spoken and written,
    Is primarily a mode of conveying ideas and emotions,
    That is, of directing and organising thought,
    As well as transmitting information.
Language is essentially a tool,
    Which allows humans to come to terms with their environment,
    And solve problems.

Language allows information which might be sophisticated or abstract,
    To be passed quickly,
    From one person to others,
    Effectively short-circuiting the need for those others to derive that information,
    From first principles.
Language also allows humans to discuss events in the past,
    Or construct theories about the future.

Once humans have learned language,
    They are able to combine words in new ways,
    To express and organise new ideas.

This ability to categorise thought,
> Allows humans to store and relate virtually limitless amounts of information,
> So that, for example, the vast array of written language,
> Represents a store of data,
> For future use by reading,
> Thus accelerating learning in those who use this data.

Individuals who have faulty language processing,
> Such as dyslexics,
> Who may have difficulty interpreting written language,
> Are often seriously inhibited from learning,
> And may suffer severely from lack of comprehensible communication.

Dyslexics may write in ways which seem confused,
> Or apparently have poor memory retention,
> And even lack of motor co-ordination,
> Which has been strongly suggested as a link with delinquent behaviour,
> Since difficulty in learning or communication,
> Can lead to frustration.

52 HOW DOES REPEATED USE OF A MOTOR SKILL RESULT IN THE ESTABLISHMENT OF A MOTOR PATHWAY?

(15)

Introduction

Learning of any kind, changes the way in which humans behave, think and feel. One example of such a change is the development of circuits in the nervous system responsible for movement.

RELEVANT POINTS

The learning of a motor skill,
    Is actually a special case of stimulus-response learning,
    Which establishes connections between sensory and motor nerve fibres,
    And causes changes within motor systems,
    By perhaps forming new synapses,
    Or strengthening existing ones,
    By some growth process,
    Or metabolic change,
    Which causes a motor nerve fibre to fire,
    When normally it would not.

Learning motor skills forms relationships between a response and a stimulus,
    Allowing humans to adjust their behaviour,
    According to the consequences of that behaviour,
    So that, for example, if the behavioural consequences are favourable,
    The behaviour pattern becomes reinforced,
    And this is reflected by anatomical changes in the nervous system,
    In connections between fibres associated with perceiving the stimulus,
    And those associated with reacting to it.

Learning a motor skill differs from other forms of learning,
    In the degree to which the new behaviours are learned,
    So that the more "innovative" the behaviour,
    The more extensive is the modification of the nerve circuits in the brain.

Establishing motor pathways by repeated use,
> Cannot occur without input,
> And subsequent changes to the sensory association areas in the cortex,
> For example, skilled movements associated with playing a musical instrument,
> Involves sensory interaction with the instrument and its sound,
> Which requires changes in sensory association areas of the visual and auditory cortices.

Even the "basic" motor skill of walking,
> Involves feedback,
> From joints,
> Stretch receptors in muscles,
> Images from the eyes.

Developing motor pathways, while involving changes in nerve circuits which control movement,
> Is therefore also heavily dependent on perceptual learning,
> Requiring changes in associated sensory nerve circuitry,
> As seen, for example, in the differing appreciation of the significance to a driver of a red traffic light,
> Compared with someone who has never driven a car.

53 DISCUSS HUMAN BEHAVIOUR UNDER THE FOLLOWING HEADINGS:
    (a) imitation; (3)
    (b) reinforcement; (5)
    (c) shaping; (4)
    (d) extinction. (3)

Introduction

Much of human behaviour is learned by imitating the behaviour of others or by having particular behaviours reinforced or shaped to a desired end. Behaviours which are not reinforced will eventually decay, a process called extinction.

RELEVANT POINTS

(a) Imitation

Imitation is the observation,
    And subsequent copying of the behaviour patterns observed in others,
    Usually directed by the individual who is imitating,
    As seen for example in an adult learning a second language,
    Allowing the individual to learn quickly and efficiently.

(b) Reinforcement

Reinforcement is the enhancing of learning,
    By using some stimulus,
    Such as an act, information or voice expression,
    Usually positive in the case of human learning,
    To reward a particular behavioural outcome,
    Thereby increasing the likelihood of that behaviour being repeated,
    As seen for example in the judicious use of praise by a teacher working with young children.

(c) Shaping

Shaping is the gradual and deliberate modification of behaviour,
    By selective reinforcement,
    Of approximations,

Which come increasingly close to a desired behavioural outcome,
And is often associated with developing complex behaviours,
By combining existing simple patterns,
As for example in the development of a particularly sophisticated athletic skill.

(d) Extinction

Extinction is the gradual decay of a particular learned response,
When the reinforcement of that response is withdrawn or reduced,
And may account for the fading of particular motor skills,
Through lack of use.
The type of reinforcement used heavily influences extinction,
With behaviour which has been learned using negative reinforcement,
Being relatively resistant to extinction,
While behaviour which has been learned using positive reinforcement,
Is much more liable to decay.

54 BY MEANS OF SUITABLE EXAMPLES, DISTINGUISH BETWEEN GENERALISATION AND DISCRIMINATION AS THESE APPLY TO HUMAN BEHAVIOUR

(15)

Introduction

It has been known for a long time that many human behaviour patterns can be conditioned. For example, a child who has to undergo regular unpleasant therapy in hospital, may associate the smell or colour of that environment when encountered elsewhere, with an irrational feeling of anxiety. An extension of this type of conditioning is generalisation and its complementary process called discrimination.

RELEVANT POINTS

Generalisation occurs when a specific behaviour pattern,
    Is evoked in response to a wide range of stimuli,
    Which are similar to each other,
    But may not all have been previously encountered,
    In the formation of the original behaviour pattern.
Generalisation is a normal phenomenon,
    Used by humans to react to new stimuli,
    Which are similar to ones already encountered,
    And is fundamental to the acquisition of knowledge and skills,
    So that, for example, a young child who is completely at ease playing with a pet cat,
    May feel similarly at ease in approaching all cats,
    While the reaction of drivers on hearing the sound of an alarm from an ambulance, police vehicle or fire engine,
    Is similar, that of giving way,
    Even though the alarms do not all sound the same.
Generalisation allows humans to predict the consequences of particular new stimuli,
    By comparing these with previous experiences involving similar stimuli.

Discrimination occurs when humans respond differently to,
    Different but related stimuli,
    So that, for example, the parents of the young child mentioned above

may act as a discriminatory stimulus,
By imposing some restrictions on the child's behaviour,
To reinforce the handling of the pet cat but not other cats,
Even if they look similar.

Discrimination is effective when the stimuli are such as to clearly distinguish,
Situations where the response should or should not be made.
Thus discrimination is a process whereby a response which has been previously learned,
May occur in more contexts than those in which the response was initially learned.

55. PEOPLE MAY SHOW ANTI-SOCIAL BEHAVIOUR WHEN WITH A GROUP BUT NOT WHEN ON THEIR OWN. HOW CAN THIS BE EXPLAINED?

(15)

Introduction

Much of human behaviour is affected by the interaction with and influence of other people. Some of these social influences can be less obvious but still elicit dramatic responses. For example, individuals who are normally law-abiding members of society, can behave quite differently in a mob situation.

RELEVANT POINTS

Groups can have powerful influences on human behaviour,
    So that it is possible for a person to lose the sense of individuality,
    When in the presence of others,
    And behave quite differently in this context,
    As compared with being alone,
    Demonstrating a process called deindividuation.
If this altered behaviour is generally deemed unacceptable,
    It is called anti-social,
    And may be the cause of many mob or crowd behaviours,
    Which result in violence and damage to society.

The conditions present in a group situation,
    Can lead certain individuals to cease to operate as separate persons,
    With little self-awareness.
Such individuals can then have,
    Very little feeling of personal responsibility,
    A sense of being unidentifiable,
    Lowered thresholds of emotional responses,
    Increased sensitivity to cues,
    Little control of personal behaviour,
    Almost no concern about the consequences of particular actions,
    Virtually no facility for rational planning,
    To become part of a potentially unruly mob.

Being in a group itself does not herald the onset of deindividuation,
    Since other variables must be present,
    As mentioned above,
    Especially the increasing feeling of personal anonymity.

56 DISCUSS THE STRATEGIES USED TO ALTER HUMAN ATTITUDES UNDER THE FOLLOWING HEADINGS:
   (a) **internalisation;** (9)
   (b) **identification.** (6)

Introduction

The concept of attitude is fraught with inexactitude. Generally it is taken to mean an internal state, such as a consciously held belief, which explains the particular behaviour of a person. Advertising companies can exploit this in marketing their products by attempting to change these internal states, that is, by altering the attitude set of the potential consumers, so that they are favourably disposed to buying those products.

RELEVANT POINTS

(a) Internalisation

Internalisation is usually taken to mean the acceptance of attitudes,
   Which may have been generated by others,
   As one's own.
Advertising companies are skilled in replacing or altering,
   Existing attitude sets towards particular products,
   By a subtle combination of well-tried strategies,
   Such as use of radio and television commercials,
   Newspaper advertisements,
   Specifically tailored to the target audience.

Internalisation takes place from early development,
   For example a child's concept of role,
   With regard to gender or age,
   Is very largely shaped by the effect of parental influence.

When the internalisation process has been particularly effective,
   Individuals may be unaware that their behaviour patterns are strongly role-based,
   Especially when in a particular environment,
   Such as the work or study area.

It has been shown that behaviour patterns based on an internalisation process are usually long-lasting,
> Often beyond the situation which actually requires that behaviour pattern.

(b) Identification

Identification occurs when a person ascribes to oneself,
> The characteristics of another person or group,
> Forming a link with that other person or group.

A child, for example, may learn a behaviour,
> By initially imitating another person,
> But gradually come to identify with that person.

It has been suggested that much of human learning about roles in society and gender,
> Occurs through this process of internalisation.

For this reason the type of "role-models" a child encounters,
> Can have a pronounced bearing on later development,
> Particularly in relation to social behaviour.

This phenomenon may be exploited for example in educating young children,
> By linking a particular life-style or value set,
> Such as abstaining from smoking and/or drugs,
> With a well-known figure in sport or music,
> Who publicly affirms the positive value of such abstentions.

57. DISCUSS HUMAN POPULATION GROWTH UNDER THE FOLLOWING HEADINGS:
(a) modern human pre-history; (7)
(b) growth to the 20th century. (8)

Introduction

It is difficult to estimate human population growth prior to about the last four or five hundred years because little documentation exists. All that can be done is to make best guesses based on archaeological and other related evidence. Nevertheless, it is possible to see that there is a huge increase in the growth rate of the human population now compared with "pre-historic" times.

RELEVANT POINTS

(a) Modern human pre-history

The study of population growth,
    Called demography,
    Requires accurate statistical records of population sizes,
    Over many centuries,
    Before reliable conclusions can be made,
    About changes in birth, death and growth rates.
Only within the last few hundred years,
    Have such records been kept,
    So that estimates of population trends prior to that time,
    Must rely on alternative evidence,
    Such as archaeology and anthropology,
    And studies into population dynamics of existing societies,
    Which follow a life-style of hunter-gatherer,
    Similar to our ancestors.

The growth rate in human populations up to about ten thousand years ago,
    The so-called "tool-making" period,
    When humans were increasingly able to control their environment,
    Has been estimated to have been of the order of 0·001%,
    For a total population size of about ten million,
    Which probably was spread over a large geographical area,
    In small groups with little contact with each other.

When humans switched life-styles from wandering hunter-gatherer,
> To one based on a stable agriculture,
> When specific crops were cultured,
> And animals domesticated,
> The "carrying capacity" of the environment increased,
> Followed by a rise in the growth rate of some one hundred fold,
> Which was maintained until relatively recent times.

(b) Growth to the 20th Century

With the sudden acceleration of industrialisation,
> In the latter part of the eighteenth century,
> The human population growth rate began to turn exponential,
> Giving a growth rate of nearly 2%.

If this growth rate persists,
> It is estimated that the human population could reach some six and a half billion,
> By the end of this century.

A number of factors have contributed to this exponential increase,
> Humans are no longer subject to predators,
> Since they now occupy the top position in many food chains,
> While factors such as infant mortality or disease,
> Have a decreasing impact,
> Improving agriculture, nutrition, sanitation,
> Medicine, education, housing or transportation,
> All increase life-expectancy and childhood survival.

The exponential increase has been called the "population explosion",
> And is a topic of great debate in projections,
> As to how the planet can support and survive such an increase.

58. BY MEANS OF SUITABLE EXAMPLES, DISCUSS FACTORS WHICH CAN CONTROL HUMAN POPULATION GROWTH

(15)

Introduction

Some parts of the world have seen a massive increase in their population during the twentieth century. This rate of increase cannot continue indefinitely since the carrying capacity of the environment is not unlimited. Like other organisms, humans are subject to factors which control population growth.

RELEVANT POINTS

Human population growth is now exponential,
    And since food and space are finite,
    This growth cannot continue indefinitely,
    Without some regulatory factor(s) operating,
    To control this trend.

Factors which control population growth,
    Are often those which control the growth at the level of the individual members,
    Such as food availability, clean water, space or disease,
    Or natural disasters,
    Such as flooding or fire.

The growth of the human population is not geographically uniform,
    And varies in different parts of the world,
    Where the rates of birth and death,
    And life expectancies,
    Vary considerably.

In poorer countries the age profile of the population,
    Is very different compared with more affluent countries,
    So that, for example, in poorer countries fewer individuals survive to reproductive age,
    Or live as long,
    As compared with richer countries.

In developing countries the rates of birth and death are both very high,
    With a high rate of infant death,
    Associated with poor hygiene,
    And nutrition.

A diet which is nutritionally deficient,
    Is known to inhibit sexual maturity,
    For a number of years,
    Thereby decreasing the potential reproductive span,
    Of individuals so affected.
A diet which is nutritionally sufficient,
    Is known to accelerate sexual maturity,
    As well as prolonging the onset of declining reproductive activity,
    Which effectively extends the potential reproductive span.
If food is scarce or nutritionally poor,
    Reproduction is known to be much less successful.

Where humans live in densely populated areas,
    Infectious diseases which might be life-threatening,
    Can spread much more easily,
    As well as mutating into new forms,
    Possibly resistant to conventional chemotherapy.

59. SCIENTIFIC ADVANCES HAVE RESULTED IN MORE EFFICIENT FOOD PRODUCTION. DISCUSS THIS STATEMENT
(15)

Introduction

Food is an essential resource and is known to be one of the most important factors regulating human growth. Much scientific effort goes into advances in techniques and strategies for growing more and better food.

RELEVANT POINTS

Improved agricultural equipment includes,
    The use of powerful petrol-driven farm machinery,
    To turn over large areas of land,
    For cultivation,
    In a relatively short time,
    Requiring less human labour,
    Or to pull sophisticated devices for planting,
    Dispersal of chemicals or harvesting.

Increasing awareness of biology of food production,
    For example by rotating crops,
    Prevents the build-up of pests,
    While deliberate planting of nitrogen-fixing plants,
    Which are rotivated back into the soil,
    Improves its texture,
    And restore the nitrogen content.

Selective breeding,
    Now allows genetic selection,
    Of plants or animals which have desirable phenotypes,
    Such as disease resistance,
    High yields,
    Desirable taste or appearance.

Intensive farming strategies,
    For growing cattle, poultry or fish,
    Obtaining eggs and milk,
    Using so-called "intensive" or "factory" farming,
    To optimise growth with minimum capital outlay.

Improved methods of storage,
    Preserve the quality of the food harvested,
    And extend its "shelf-life",
    By inhibiting the growth of decomposers,
    Using low temperatures.

Control of weeds and pests,
    Using a wide range of highly specific chemicals,
    Can target quite precisely particular weeds or pests.

Fertilisers, either organic or inorganic allow,
    Soils to be intensively cultivated,
    By maintaining their nitrogen, phosphorus and potassium content,
    Which is removed by growing high-yield crops,
    As well as improving the soil texture,
    Reducing erosion.

60 WHAT ARE SOME OF THE PROBLEMS OF PROVIDING CLEAN WATER AND HOW CAN THESE BE SOLVED?

(15)

Introduction

Humans can survive for only a short time without water which is intimately bound up with life itself. Good health is heavily dependent on a supply of clean water and increasing demands worldwide generate a need for sophisticated techniques to ensure quality is maintained.

RELEVANT POINTS

Some 1% of the earth's surface water is actually fresh,
    But may not be suitable for direct drinking.
The unsuitability of water may be due to contamination by chemicals,
    Such as aluminium,
        Which is thought to be linked to the incidence of Alzheimer's disease,
    Or lead,
        Which is now known to be associated with brain damage and mental retardation in children,
    Or nitrates,
        Which may be carcinogenic.
Run-off water from land which has been treated with pesticides or herbicides,
    May contain dangerously high levels of compounds,
        Such as simazine.
Water from industrial processing,
    For example, dry-cleaning and paint manufacturing,
    May contain solvents,
        Such as chloroform and trichloroethylene.
Another source of contamination may be microbes,
    Causing water-borne diseases,
        Such as cholera, typhoid, polio and hepatitis.
Water can also provide a home for the vectors of diseases.
    For example, the larvae of the mosquito,
        Which may spread malaria.

In order to render water safe to drink,
> It must be treated to remove possible harmful ingredients,
> Using a combination of techniques such as filtration, coagulation and chlorination,
> Then safely stored and transported to where it is required,
> Without loss of quality.

Regular well-tried bacteriological examinations are made,
> To monitor the presence of particular indicator organisms,
> Such as *Escherichia coli*,
> A known inhabitant of the human gut.

The increasing demand for water puts pressure on existing provisions,
> Which require to be constantly maintained and upgraded.

The provision of clean healthy water requires considerable social co-operation,
> Which is influenced by local conditions, financial constraints and cultural attitudes.

The United Nations, for example, has agreed international standards of water quality,
> Which depend on a battery of tests,
> Meeting stringent standards.

61 DISCUSS DISEASE UNDER THE FOLLOWING HEADINGS:
(a) regulatory effects on populations; (8)
(b) use of vaccines. (7)

Introduction

Human populations are subject to the same biological regulatory factors, including disease, as other animals. By the use of appropriate vaccines, it is now possible to reduce or eliminate entirely the effects of these diseases.

RELEVANT POINTS

(a) Regulatory effects on populations

The regulatory effects of disease are not evenly spread,
    But are closely linked with geographical location,
        Which brings in variables such as climate, cultural attitudes, available wealth, living conditions, standards of hygiene or nutrition,
        So that, in general, underdeveloped countries are often most affected.
The control of infectious diseases,
    Has been a main contributory factor,
    In the rapid rise of human populations,
    Since the Second World War.

Disease in humans may be caused by pathogenic organisms,
    Such as bacteria, viruses, fungi or protozoa.
If unchecked, pathogens can exert considerable regulatory effects on human populations,
    Perhaps by debilitating the human host,
    Reducing the ability to successfully reproduce,
    Or by causing actual death.
Most unchecked pathogens regulate human populations,
    By reducing the rate of offspring production,
    And/or increasing the rate of mortality.
Diseases may be tolerably present at a low level in many human populations,
    Causing relatively little damage,
    Unless the numbers and spread of infected persons exceed some threshold,
    When epidemics can occur,

Affecting many individuals,
Especially if these individuals are in close proximity.

(b) Use of vaccines

Knowledge of the causative agents of infectious diseases has increased dramatically in relatively recent times.
Such knowledge has enabled the construction of strategies for controlling infectious diseases,
  Using vaccination programmes,
  Which seek to prevent, rather than cure, disease,
  By deliberately stimulating the body's own immune system,
  To actively produce highly specific antibodies against the pathogens.
Diseases which were at one time labelled "killers",
  Such as typhoid, diphtheria, polio, tuberculosis or smallpox,
  Can now be prevented by the use of vaccines.
Smallpox epidemics,
  Which devastated populations in the past,
  No longer occur.

In highly developed countries the level of protection by vaccination,
  Is of the order of 90%,
  But much less in underdeveloped countries.

62 DISCUSS THE EFFECT OF POPULATION GROWTH ON THE DISRUPTION OF FOOD WEBS

(15)

Introduction

A food web is a complex mix of a number of food chains linked together. Factors which affect any one element of a food web can have profound consequences. One such factor is increasing population growth.

RELEVANT POINTS

Early humans probably survived by hunting and gathering food,
    As members of a balanced ecosystem,
    In which their numbers were kept relatively stable,
    By factors such as predation, disease or inadequate food supplies.
As the pattern of life changed to one based on agriculture,
    In which crops were grown and harvested,
    And animals were domesticated,
    Population growth increased dramatically,
    And began to have a significant impact on the environment,
    With consequent effects on balanced food webs.
These effects have expanded enormously in relatively recent times,
    With parallel developments in medicine, improvements in nutrition and hygiene,
    So that the human population is growing exponentially,
    And physically requires more space to survive.

The utilisation of land nearly always means the loss of potential growing areas,
    So that food production has to be maximised and efficient,
    To support the increasing population.
This food production often involves the use of chemicals,
    Such as fertilisers, herbicides or pesticides,
    Which can lead to pollution problems,
    In waterways,
    As well as accumulating in food chains,
    Damaging soil structure,
    And inevitably harming organisms,

Other than the target species,
Thus affecting balanced food webs.

Associated with increasing human population growth is an upward trend in industrialisation,
Such as the development of factories, chemical plants and refineries,
Which may discharge noxious chemicals,
Such as mercury, lead, cyanide into the water,
Killing aquatic life and reducing aquatic oxygen levels,
Causing possible long-term damage to food webs.

The enormous rise in the use of fossil-fuel technology,
Has also seen a huge increase in the amount of atmospheric pollutants,
Which have been shown to have a wide variety of effects on food webs,
Such as plant damage.
Since almost half the human population live in towns,
There is a necessary demand for efficient travel,
Which may involve transport schemes which effectively cut across natural habitats,
By roads or rail.
Dividing balanced ecosystems into smaller units,
Which may be unable to survive.

The greater the number of people there are,
The more sewage is produced.
Food webs in streams and rivers can be seriously damaged,
By the discharge of sewage, even when treated,
Due to the raising of nutrient levels,
Causing eutrophication,
And dramatic changes in the aquatic wildlife.

With the development of sophisticated and powerful technology,
Large areas of land can now be cleared quickly,
For monoculture,
Or the building of houses,
Irreversibly altering associated food webs,
As well as possibly encouraging soil erosion.

The alteration of the path of natural rivers,
> To be used in irrigation schemes in some parts of the world,
> Can cause salt depositions on the surface,
> Which causes the soil below to have an abnormally high salinity,
> Changing considerably the natural habitat for organisms living there.

Another effect of irrigation schemes is the leaching of toxic chemicals,
> Such as selenium from rocks into the environment,
> Which can have a devastating effect on food webs.

63. DISRUPTION OF ESSENTIAL NUTRIENT CYCLES CAN HAVE SERIOUS ENVIRONMENTAL CONSEQUENCES. DISCUSS THIS STATEMENT WITH RESPECT TO THE NITROGEN CYCLE (15)

Introduction

Nitrogen is a vital element for growth and is recycled constantly within balanced ecosystems. Green plants uptake the nitrogen and incorporate it into compounds such as proteins and nucleic acids. When the plants die or are eaten, the nitrogen is released to be used again. Since nitrogen is so important, if the cycle is damaged or interrupted, the environmental consequences can be serious.

RELEVANT POINTS

Two processes are involved in the natural cycling of nitrogen,
- Immobilisation,
- In which inorganic nitrogen is accumulated as organic nitrogen,
- And mineralisation,
- Which is the reverse.

Both these processes involve soil microbes,
- Which may be affected by a number of factors,
- Such as soil water content,
- Oxygen levels,
- Temperature,
- And pH.

If natural microbial activity is altered,
- For example, as an indirect consequence of using a pesticide,
- Such that the conversion of nitrite to nitrate was inhibited,
- This might cause toxic levels of nitrite in drinking water.

Any activity which raises the nitrogen level in the environment,
- May have serious consequences on the ability of the nitrogen cycle to assimilate the increasing load.

Where soil fertility is poor,
> This is often supplemented by the use of chemical fertilisers,
> Rich in ammonium nitrate.

If these chemicals are leached out by rain,
> Into rivers and lakes,
> The effect can be to artificially raise the nitrogen level,
> So that aquatic plants and algae in the short-term,
> Increase in numbers dramatically,
> Causing so-called "algal blooms",
> Or eutrophication,
> Which can kill large numbers of aquatic organisms,
> And seriously deplete the oxygen content of the water.

Sewage is extremely rich in nitrogen compounds,
> And if discharged raw or improperly treated,
> Can have an effect analogous to that of leaching of fertilisers,
> Since the nutrient level of the water would be raised,
> Causing microbial growth,
> Reducing the oxygen content,
> And releasing atmospheric ammonia.

Even after treatment sewage still contains nearly half its original nitrogen content.

The burning of fossil-fuels,
> Causes the formation of harmful nitrogen compounds,
> Such as nitrogen oxides,
> Which are discharged into the atmosphere,
> And may cause photochemical pollution,
> When strong sunlight catalyses chemical changes,
> In the nitrogen-containing emissions,
> To form nitrogen dioxide and peroxyacyl nitrates,
> Both of which have been shown to have a harmful effect on plant life.

64. DISCUSS SOME OF THE POSSIBLE DETRIMENTAL EFFECTS ON THE ENVIRONMENT OF INCREASING THE LEVELS OF ATMOSPHERIC CARBON COMPOUNDS
(15)

Introduction

The atmosphere may be regarded as a vast reservoir of carbon dioxide which photosynthetic organisms can use. This fixed carbon is then available to other organisms as a source of energy in the process of respiration. The balance of this natural cycle can be altered by human activity and have world-wide effects such as global warming and even changes in the climate.

RELEVANT POINTS

It has been estimated that the burning of fossil fuels,
    Is adding carbon dioxide to the atmosphere,
    At the rate of some 2·3 parts per million per year.
Most of the increased carbon dioxide production is absorbed,
    Perhaps by dissolving in ocean waters,
    Which act as a "sink",
    Or being utilised by increased photosynthetic activity.
Not all the extra carbon dioxide can be absorbed,
    Which has a displacing effect,
    On the equilibrium of the carbon cycles.

The natural land and sea carbon cycles,
    Can cope with most but not all of this additional loading,
    And the concern is that the increasing carbon dioxide levels,
    Could herald the onset of global warming,
    Causing polar ice caps to melt,
    Sea levels to rise,
    Flooding of rivers,
    And changing weather patterns.

Carbon dioxide allows solar short-wave radiation to pass through easily,
    But absorbs long-wave radiation reflected back from the earth's surface,
    Causing a rise in the temperature of the lower atmosphere,
    Otherwise known as the "greenhouse effect".

A similar effect is caused by other carbon compounds,
  Such as carbon monoxide,
  From car exhaust fumes,
  Chlorofluorocarbons (CFC gases),
  Used as "inert" propellants,
  And methane,
  Which is a natural product of decomposition,
  But increasingly released from combustion processes,
  Or leaks from gas distribution systems.

Present knowledge of global meteorology,
  Cannot allow accurate predictions of the effects of global warming,
  Which supports the view that continued damaging of the natural carbon cycle,
  Should actively be discouraged.